PROFESSIONAL ENGINEERING PRACTICE ETHICAL ASPECTS

CONTRIBUTORS

George Adam, P.Eng.
Special Consultant
Con-Force Structures Limited
Calgary, Alberta
(Chapter 6)

Hugh Adcock, P.Eng.
Assistant Deputy Minister, Retired
Ministry of Transportation and Communications
Government of Ontario
(Chapter 4)

Howard Bexon, P.Eng.
Consultant, Retired
Material Handling Engineering Department
Dofasco Inc.
Hamilton, Ontario
(Chapter 5)

Oswald Hall, Ph.D.
Professor Emeritus
Department of Sociology
University of Toronto
(Chapter 2)

John Kean, P.Eng.
President
Canadian Standards Association
(Chapter 7)

George Sinclair, Ph.D., P.Eng.
Chairman of the Board
Sinclair Radio Laboratories Limited
Aurora, Ontario
(Chapter 10)

John Stevens, P.Eng.
Atomic Energy of Canada Limited, Retired
President, Strangeways-Stevens
Consulting Engineers
Rexdale, Ontario
(Chapter 8 and a case study in Chapter 12)

PROFESSIONAL ENGINEERING PRACTICE ETHICAL ASPECTS

Carson Morrison, P.Eng.
Professor Emeritus
Civil Engineering
University of Toronto

Philip Hughes, P.Eng.
Professor Emeritus
Mechanical Engineering
University of Toronto

Third Edition

McGraw-Hill Ryerson Limited
Toronto Montréal New York Auckland Bogotá
Caracas Lisbon London Madrid Mexico
Milan New Delhi Paris San Juan Singapore
Sydney Tokyo

PROFESSIONAL ENGINEERING PRACTICE: ETHICAL ASPECTS
Third Edition

ISBN: 0-07-551284-X

1 2 3 4 5 6 7 8 9 10 D 0 9 8 7 6 5 4 3 2 1

Printed and bound in Canada

SPONSORING EDITOR: CATHERINE A. O'TOOLE
SUPERVISING EDITOR: MARGARET HENDERSON
PRINTING & BINDING: JOHN DEYELL COMPANY
TEXT SET IN: TIMES ROMAN

Canadian Cataloguing in Publication Data

Morrison, Carson, date-
Professional engineering practice : ethical aspects

3rd ed.
ISBN 0-07-551284-X

1. Engineering ethics. I. Hughes, Philip.
II. Title.

TA157.M67 1991 174'.962'0971 C91-094773-2

TABLE OF CONTENTS

PREFACE

Throughout history the community has been concerned with the manners of its members: it has ruled what is, and what is not, acceptable behaviour. Our community of professional engineers in Canada has laws, statutes, regulations and codes to define what is, and what is not, acceptable behaviour in our community.

In this work we are concerned with those laws, rules, regulations and codes that are relevant to the practice of professional engineering in Canada's provinces and territories, and the related practice of geologists and geophysicists in Alberta and in the Northwest Territories. We consider and comment on the professional engineers' Codes of Ethics in the provinces and territories, and the Code de déontologie in Quebec.

Essentially, the engineering profession in Canada—with some variations in details and mechanisms—is self-regulating under the Acts which, along with certain by-laws and regulations, are intended to ensure orderly and systematic conduct in relations between the profession and the public. Codes of Ethics and other regulations govern the practice of the individual practitioner.

Regulation of engineers and other professionals in Canada is the responsibility of the provinces and the territories. The main objectives of the Acts or Ordinances that relate to professional engineering practice in each of Canada's 10 provinces and two territories are similar but there are differences in some of the details. A practitioner should refer to the relevant legal instruments in any jurisdiction where he practises.

In each of the jurisdictions, professional engineering has the status of a self-regulating profession and, notwithstanding some differences, there is an overall similarity between the self-regulating rules and procedures in all of them.

The Ontario Professional Engineers Act, Bill 123, Chapter 13 of the Statutes of Ontario 1984 and O. Reg. 538/84 made under that

Act have brought into effect significant changes in the regulation of the profession in Ontario.

The objective of this book is to enhance the understanding by an applicant for membership in any of the 12 associations of professional engineers in Canada's provinces and territories, of the ethical obligations of a member.

Each association has its own Code or Canons of Ethics. An overall similarity and harmony of intent pervades these codes, yet there is sufficient difference in emphasis and in wording and arrangement to warrant the inclusion of all 12 codes. Not only does it make the book more widely appropriate, but for candidates it illustrates and emphasizes that they share a common heritage of ethical principles, as they do of education and training, with Canadian engineers in all jurisdictions.

The Ontario Professional Engineers Act came into force September 1, 1984. Section 1.(m) of the Act states that: "the practice of professional engineering" means any act of designing, composing, evaluating, advising, reporting, directing or supervising wherein the safeguarding of life, health, property or the public welfare is concerned and that requires the application of engineering principles, but does not include practising as a natural scientist.

The authors acknowledge the kind permission of:

1) The professional engineering associations (the Ordre in Québec) in all of Canada's 10 provinces and two territories, to use their codes of ethics;
2) The Institute of Public Administration of Canada, to use excerpts from its Ethical Guidelines for Government Employees by Kenneth Kernaghan;
3) The Royal Bank of Canada, to use the *Royal Bank Letter*, Vol. 62, No. 1, Jan./Feb. 1981.
4) Alan Bernard, P.Eng., Deputy Registrar of APEGGA to use information obtained from him to describe the role of the geologist, geophysicist and engineer in the Resource Industry.

CHAPTER 1

DEFINITIONS

A clear understanding of the meanings of the key words relevant to the subject is essential to an understanding of the ethical aspects of professional engineering practice. For that reason, this chapter lists, defines and comments on the meanings, origins and usage of some of these words.

Definitions are selected from the *Oxford Universal Dictionary*, third edition, 1955; several quotations are taken from Fowler's *A Dictionary of Modern English Usage* (second edition, 1965). Definitions and quotations represent points of anchorage upon which the authors have hung a discursive context.

An early recognition of the professional engineer appears in Adam Smith's *The Wealth of Nations*, 1776, where in pursuing the subject of the division of labour, the author speaks of

> . . . those who are called philosophers, or men of speculation, whose trade it is not do any thing, but to observe every thing, and who, upon that account, are often capable of combining together the powers of the most distant and dissimilar objects.

It is doubtful that this quaint description could be improved two centuries later; if we try to render it in modern and precise terms we encounter the same difficulty as when we try rewriting familiar Biblical passages in today's idiom with the hope that a clearer understanding may be effected: too often the pith of the original is lost, and the meaning may be limited in scope by the very precision we have attempted.

We might complain of Adam Smith's job description, "whose trade it is not to do any thing," on the grounds that the professional engineer today is likely to be run off his feet with duties, until we recall that Adam Smith's contemporaries—the giant Watt and the formidable characters whose bridges, canals and harbour works foreshadowed and patterned our accomplishments—were half craftsmen, half philosophers, and that they exhibited an enormous capacity for work in their lifetimes. But they did not produce goods for commerce. Their accomplishments loom to this day and remain our pride and our envy, but were not the "doing of things" in Adam Smith's context of the division of labour.

Engine; Engineer

The dictionary tells us that the word "engine" is derived from the Latin "ingenium" (and thus obviously related to "ingenious" and, a connection less obvious but no less interesting, to "engender" or beget). The word is given a series of meanings. It was in use before the steam engine; its original meaning was "an instance or product of ingenuity." Likewise, the word "engineer" preceded the steam engine, and in its early use was attached to one who contrives, designs or invents; an inventor, a plotter. We now confine "engine" to the engine as we know it and other meanings are seldom encountered. "Engineer," however, when we couple to it the adjective "professional," has retained its original meaning of one who contrives, designs, et cetera, because the professional engineer is by no means always associated with the engine.

The Operator Engineer

To-day, those most intimately concerned with the engine are stationary, railway, and other operating engineers, who do not formally share a duty that could be described as ". . . not to do any thing, but to observe every thing," and to whom we deny the designation professional, which we have arrogated to ourselves. Observe every thing they may not, but observe their engines they certainly do, and if our philosopher or man of speculation has added wisdom to his faculties, he will seek the benefit of the observations of these half-brothers, and accord them proper respect. Indeed, if our philosopher has occasion to cling to a walkway rail in the engine room of a ship in a heavy sea, or to watch the flight engineer vigilant before the array of instruments that monitor the great engines that are shouting the glory of God at 50 000 pounds thrust, the sense of respect will be strong in him.

Profess; Profession; Vocation

In its widest sense, "to profess" is to declare openly, to affirm before the world some loyalty, belief, adherence to a school of thought, or to claim openly the command of some field of knowledge. That field of knowledge is the profession. A narrower meaning of the word is that of the calling or vocation by which a livelihood is earned, without, at least in its use in the past, a particular connotation of learning. Our dictionary gives us an example of usage: "Joseph, her spouse, by profession a carpenter." This seems to suggest English usage early in the seventeenth century.

The "Learned Professions"

Common today is this definition of "profession": a vocation in which a professed knowledge of some department of learning is used in its application to the affairs of others, or in the practice of an art founded upon it.

A somewhat different, but not conflicting, definition of "profession" comes from Webster's *New World Dictionary* (Second College Edition, 1970): a vocation or occupation requiring advanced education and training and involving intellectual skills, as medicine, law, theology, engineering, teaching, et cetera.

From before the Industrial Revolution, say about 1700, to the late years of the nineteenth century we spoke of three learned professions—divinity, law and medicine—and of two not associated with learning—the military profession and the profession of gentleman (!). These thrived or starved as the case might be for the individual; all five enjoyed an irreproachable respectability in the middle class of society.

Today's Professional

Gone is the public consensus that so limited the professions. We seldom use the tag "learned" now, but to apply to a calling the designation "profession" we tend to require that there be a "professed knowledge of some department of learning" to justify it. This, used as a criterion, gives us no direction as to what is to be considered acceptable as a department of learning; as with the mathematical model that shows what critical decision has to be made, it leaves the rest to us. In fact, except as we are members of the public, we do not make any rule or decision in the matter, for it is public repute, the public image, that decides whether a calling is a profession in the eyes of the world.

Do We Care?

Yes, we do care. We believe the learning we profess and apply is beyond question adequate, and we believe the engineering profession has been accepted by the public for many years now. But our caring goes deeper than concern with public regard, goes into other and more subtle aspects of professionalism, into the satisfaction and happiness each of us, as an individual, derives from his practice.

For years, texts on management have given us lists of human satisfactions for which provision should be made in planning industrial working conditions. But for the individual practitioner of the beneficent and terrible art we profess, let us look beyond the relatively obvious range of the working place. In every man and woman there is a flicker of desire, most of the time subconscious, to serve some end quite beyond personal welfare, beyond the need to be admired, to be loved, to be procreator of future generations. Indeed, this may be the presence of the abstract: love. And happy, if seldom prosperous, are the few in whose lives this is the dynamic, who might say, as the father of the famous Mayo brothers is said to have said, "There is no greater disgrace than to die rich."

We number few saints among us. But as we achieve command of the "department of learning" that is the province of engineering, that faint flame may reveal a further knowledge: that by venturing into this profession, we have pledged ourselves, perhaps inadvertently, not only to public safety as it is affected by our works, but to grave concern with the quality of the future of our kind. And if your authors have adverted to imaginative language in expressing these solemn thoughts, it is an attempt to crack the shell of superficiality in which the meaning of the word "professionalism" is so often encased, an attempt to capture something of the notion of ordination that is implicit in the term; it is a summons to "think big" about what we do.

Ethics and Morality

These words are both defined and differentiated in one golden sentence given us by Fowler: "Ethics is the science of morals and morals are the practice of ethics."

Let us put Fowler's sentence in a slightly more explanatory cast: Ethics is the science by which people draw up sets of moral imperatives; the morals of an individual are the way in which he or she applies such a set of imperatives in his or her actions.

Note that the word "ethics," written and spoken with the final *s*, is a singular noun. A prevailing code of morality is one of the meanings given in the dictionary, and is the closest to what we generally intend by the word.

We encounter no less than 23 square inches of small print in the defining of "moral" as a noun, adjective and even verb (rare), with "morals" as the plural noun. From this vast array we boldly extract for "morals": "character or disposition; distinction between right and wrong, or good and evil, in relation to action."

The word "ethic," without the final *s*, is properly used as an adjective, and is synonymous with the word "ethical." As a noun, it means ethics; it is not properly used to refer to one canon of a group, or set of canons. For instance, the Protestant ethic, sometimes cited in management texts, refers to the code of ethics of Protestantism, not to a particular part of such a code.

Sometimes the negative, or converse, forms of words have peculiar shades of meaning. Fowler says that since the word "immoral" is popularly associated with sexual immorality, the word "unethical" has come into vogue as an adjective that describes the person who is immoral in other ways, especially in violating the accepted code of a profession or business. The words "right" and "wrong" and "good" and "wicked" seem now to be old-fashioned, according to Fowler.

A previous mention of the 23 square inches of small print used to define a single word was intended to convey an idea of your authors' diligence; applying it they have concluded that the usage adopted by the Engineers' Council for Professional Development (ECPD) is, as might be expected, rigorous. ECPD speaks of its "revised set of Canons of Ethics (1973)." A code of ethics is a set of canons. We can speak of a certain canon in a code without having to use words like "rule" or "precept," neither of which has quite the same meaning as canon.

Associated historically with ethics is Hippocrates of Alexandria, third century B.C., whose writings about medicine are said to deal explicitly with ethical aspects of that art—too explicitly for modern practice, perhaps, for the Hippocratic Oath (see under "Medicine—History of" in the *Encyclopaedia Brittanica*, 1946, Volume 15), tacitly or openly avowed by medical practitioners for many centuries, has been abandoned in some parts of the world in recent years.

Codes of Ethics, Written and Unwritten

In all walks of life, codes of ethics appear, drawn up to express the "character or disposition in relation to action" of a group of persons of common vocation or interest. The extent to which the code is observed by the members of the group, and probably the extent to which it is taken seriously by the public, are both subject to much the same variables: apart from the strong factors of tradition and recognition of the trade, profession or common interest as an entity in the life of the community, an obvious factor is the authority of the group vis-à-vis its members. In a self-regulating profession like ours, a member can be struck from the professional register in cases of grossly unethical conduct, but many groups, although the members subscribe to a code, have only the withdrawal of esteem or companionship, or of a reserved title with which to enforce it. An interesting and almost converse case is that of business groups, often loosely knit, which, having no code in writing, yet wield a very real power to enforce compliance with one or more moral principles. In commerce and finance, contracts that often involve huge sums of money are made by word of mouth. That prime canon, that a man's word shall be as good as his bond, obtains without question. And the enforcement is absolute: to transgress this single canon is to be excluded from high trading circles without forgiveness this side of the grave. Remembering this helps us avoid complacency when we contemplate the written code we subscribe to in our chosen profession.

Status; Esteem; Image

We are frequently reminded of our status as members of the engineering profession, and we are expected to behave in a manner appropriate to that status. The word "status" is defined as "legal standing as determined by membership of some class of persons or things enjoying certain rights or subject to certain limitations." Let us dismiss *things* by citing an example given in the dictionary, the status of enemy merchant vessels, and consider *persons*. Our own status is that common to all citizens, except that we enjoy certain additional rights (in Canada) conferred upon us by the Professional Engineers Acts. That is, we have a particular status that is assumed by the individual upon registration. It is clearly defined; we hold it under the law so long as our name stands on the register.

Status, of itself, implies nothing about public regard. A tinker doing 30 days for being drunk and disorderly has a status, just as does an archbishop; the tinker is subject to certain limitations while the archbishop enjoys certain rights, but the condition of each fits the definition of status. The esteem, however, in which each status is held is very different.

The word "esteem" has several definitions, such as "to evaluate," as a verb, or "valuation," as a noun, the noun being further described as favourable opinion, regard, respect, which is the meaning we intend here.

Persons who hold a recognizable status, which they themselves esteem, tend to constrain themselves to act their part in the world so as to communicate that esteem to others. It is a human trait, a natural desire. Further, such persons, jealous of the esteem they cultivate in others, will resent a fellow member of the group who will not be subject to the same constraints, holding such person to be one who destroys what they strive to build. In such a sequence, the idea of a common image possesses the group, and often, as our own Associations have done in the past, the group discusses and publishes among its members various ways to achieve it.

With some care we have differentiated *status* and *esteem*, the first susceptible to measurement and the second not, but no less important. We have met, and already employed, the word "image"; we know how it is used, perhaps overused, today. The dictionary emphasizes the element of artificiality, not only by giving us the Latin root, which is common with that of the verb "to imitate," but in the definition itself.

"Image" is the artificial representation of a person; a mental picture or impression; an idea, conception. We cannot depend upon nature; each of us must design his image to be the picture he wants to appear in the minds of others.

Most of the discussions on the subject published among our own members had to do with participation in public affairs, activity in good causes, with being good citizens, and so on. The thesis addressed was the entirely rational one that popularity can, and should, enhance esteem. But there was no direct aim at the particular rights of our status as affecting the design of the image we were to seek to project. Our Codes of Ethics are directly concerned with our behaviour in the practice reserved to us, and give full attention to the public to the extent that our works affect it. However, the

Codes do not normally reach the public and, even if they did, they are somewhat remote from the opinion-forming contacts that create the public image.

Perhaps, for each one of us, it boils down to our minimizing the artificiality by being, and being seen to be, trustworthy in all matters outside our practice, just as we are trustworthy, and are seen to be trustworthy, in working relations with our fellow engineers. The good citizenship aspects we can design to suit our capabilities and desires.

CHAPTER 2

THE PROFESSIONS IN PRESENT-DAY SOCIETY

OSWALD HALL, PH.D.

In present-day society professions are very numerous. This was not always the case. Historically the professions were three: medicine, law and the priesthood. Professionals looked after one's health, one's fortune, and one's soul.

In each case they dealt with a fearfully consequential set of affairs. One's health is continually threatened by unknown dangers. One's affairs are at the mercy of the law, which seems infinitely mysterious. The fate of one's soul has been a source of endless anxiety. The three great historic professions attended to the central problems of the human animal.

Because they dealt with matters of transcending importance, these professions have been accorded a great deal of autonomy and freedom to go about their work. Historically their advice, and their orders, were unquestioned. They have been called the "free independent" professions, because they were free from control by any other part of society. It is small wonder that they have been admired and envied by other occupations, or that other occupations have tried to copy them. In all societies they have ranked at the top of the social order as far as income and prestige are concerned. Most occupations that aspire to be professions tend to model themselves on the historical professions.

In present-day society one can count the professions by the score. The later ones do not seem to provide such important services as did the three mentioned earlier. Dentistry is a fairly recent profession, and it provides a distinctive service. But dentistry is not a life-and-death sort of enterprise. However, it does seem that in the course of time new professions do arise, generally because society comes to accept their services as important. In our day the condition of our teeth, and the accompanying smile, seem almost as important as the condition of our souls.

On the other hand, many of the newer professions do not provide a service to the individual person. They provide services for organizations. In our society, organizations have an importance comparable to that of persons. In a very real sense one can say that there are professions that look after the health of organizations (here accountants come to mind); there are others that look after their fortunes (economists, for one); and if the organization has a "soul" the public-relations specialist attends to its welfare.

Science and the Professions: The Case of Engineering

If we consider engineering as a profession it is clear that the "services" are rarely provided for a person. Bridges are built for railways, skyscrapers are built for companies, oil wells are drilled for corporations, mines are dug for business concerns, and so the list goes on. In modern societies the services performed for organizations rank in importance with those provided for individual persons.

Nonetheless, engineering shares many of the earmarks of the historic professions, among which one stands out as deserving special consideration. This is the relation to science. Sociologists argue as to which is more consequential for modern society, the work of the scientist or that of the professions. Clearly each has transformed the nature of work. Clearly the two developments are interwoven in an intricate way. Their fortunes and their futures are intertwined.

On the other hand, in any particular profession, they are rarely in step. Sometimes science outruns technology, and sometimes technology leads science. In mechanical engineering the steam engine was widely used long before scientists understood the principles involved in the field of energy. By contrast in electrical engineering the nature of electricity was discovered long before anyone contrived any use for it.

One of the lively aspects of any profession is the way it deals with the growth of scientific knowledge. If it ignores science, as the priesthood has done in various times and places, a profession may decline or even disappear. At the other extreme, if a profession tries to use the full resources of science it may have great difficulty handling practical problems. One thinks of the painful condition of the engineers who seek ways to deal with nuclear wastes, and with the responsibilities of atomic explosions. Such engineers may be looked on as the "enemies" of society, rather than the providers of important professional services.

While the problem seems distinctive when stated as above, one can see that we have here a very general feature of all professions. It is the matter of the relation of the profession to the body of scientific knowledge. For any profession we can ask the following questions:

> Does the profession make use of a substantial body of scientific knowledge, or a slender body?
>
> Does the profession have procedures that keep its practising members abreast of the changes in scientific knowledge?
>
> Are the members of the profession themselves involved in research, that is, in the actual creation of new knowledge, or does it relinquish this task to others?
>
> Are the members of the profession on the forefront of the research enterprise, or are they content with modest efforts in this respect?

These four questions are, in a sense, measures of the welfare of any profession. The members of each profession need to look at them, fairly and squarely, as a guide to understanding where they are heading.

The Social Control of the Professions

As noted, the professions have striven to become "free and independent." At one level this has meant being "free" to charge substantial fees for their services, and hence to become financially independent. But in a more important way "free and independent" meant the right to decide what should be done for a client and how it should be done. In essence this means that the professional decides what is right for the client to do. In effect the professional assures society that the client is being advised in the most appropriate way. Wherever this occurs the professional becomes the final arbiter of what is good for society. Without doubt this is a tremendous privilege that is granted to the profession. The other side of the coin is the awesome responsibility it involves. Hence one can say that the professions in society enjoy very high prestige insofar as they have been granted privileges of this order and have shouldered the responsibilities that go with them.

Generally, professionals are a bit more eager to have the privileges of their way of life than they are to accept the responsibilities. On the other hand society is more prone to limit the privileges and to insist that professionals live up to their responsibilities. Hence there is a continuous tension between the professions and the larger society, and a continuous effort by society to limit the autonomy of professions.

This tension arises for two reasons. On the one hand more and more persons in our society are expecting and demanding more and better services from the professions. They are demanding more from the professions in terms of the quantity and quality of their services. At the same time organizations are expecting more from their professionals, and laying more responsibilities on them. This is so because professional failures have terrible consequences for organizations. If a dam bursts or a plane crashes, if a bridge collapses or a factory explodes or a fire-protection system fails, the effects are felt not by one individual person, but by possibly a multitude of persons. Hence society continually seeks new ways to ensure that professionals live up to responsibilities, and these responsibilities themselves become ever more inclusive and more detailed.

The efforts to control professions run along three distinct paths. The first path is to create a legal code. In this case the state, or some branch of government, establishes rules governing the particular profession. It does this by legislation, by an act of the legislature that sets out in legal terms what are the responsibilities and privileges, the rights and the duties, of the members of the profession. Such an act requires that there be a set of inspectors who regularly inspect the ongoing work of the professionals, to assure that the quality and quantity of their work is adequate. Besides these, there are legal penalties set forth and applied to those professionals whose work is insufficient in the eyes of the inspectors. Altogether these represent a system that assures society that professional services can be trusted.

The second path to social control is reflected in a code of ethics. In this case the procedures for control are vested with the profession itself, that is, the profession viewed as a collectivity. In this case the members of a profession set up some sort of association, which in its own right establishes the rules and regulations under which its members work. Such an association may establish the scale of fees and the responsibilities of its members. It may also set up machinery to check the work of its members, and ways to punish those who fail to adhere to the code of ethics. But those who do this are members of the profession, and not the police and the judges of the larger society. To say that a profession has professional autonomy means that it handles the matters just discussed.

The third form of social control can be described as a code of honour. This is not a written document like a legal code or a code of ethics. Rather it is something internal, as is the case with one's conscience. In many professions one finds a more highly developed sense of professional honour, which guides members in what they should and should not do. A code of honour works without any need of inspectors, or of police to enforce it. Such a code may demand higher standards of people than colleagues would expect. Yet it may be a source of deep professional pride to be able to live up to a code of honour.

These are the three basic forms of social control. The first is exercised by an agency outside of the profession. The second is made effective by the collective efforts of all the members of the profession. The third operates within the individual member of the profession, even though it has been learned from other colleagues. It would seem that, in order for the members of a profession to live up to their responsibilities, one or more of the three codes mentioned must be functioning.

Professional Associations

Almost all the professions have established one or more associations, which take care of the collective interests of the profession. Some professions have one association that handles many functions, while others have a variety of associations that are more specialized. It should be noted that there is usually one compulsory association whereas the others tend to be voluntary.

The medical profession has a very elaborate set of associations. Over the past century six very different types have emerged. One is called a "College," and it is compulsory that all physicians join it. This one is established by law. It controls the licensing of the members, and is responsible for their registration. It has substantial responsibilities for education and for the examination of candidates. Another of its main functions is the formulation of a code of ethics and its enforcement.

A second association is largely economic in character. It sets the fee schedule for the members, and bargains with the government regarding charges therein. In effect it is the political arm of the profession. Unlike the College, it is a voluntary association, though almost all members belong to it as well as to the College.

There is a third association that caters to professional self-improvement, and also provides opportunities for sociability for the members. In it members come together to listen to lectures and join in discussion. This association has both a library and a museum that contribute to the intellectual life of the profession. They help the members to stay abreast of developments in the field of science.

Along with these three there is another that concentrates on the concerns of the women members of the profession, while a fifth looks after the special interests of students and the doctors just starting on their careers, the internes.

Finally, there are smaller associations made up of the doctors who are associated with a specific hospital. They come together to handle the various problems that arise in the daily working of a modern hospital, particularly as these problems bear on the welfare and working conditions of doctors.

It should be noted that each of these various forms of association handles a very consequential part of the affairs of the medical profession. As noted above some professions manage to survive with only one association. In those cases, of course, the tasks are all placed on the shoulders of one executive group. In general, the more highly developed a profession becomes the more specialized associations will it develop.

To an increasing degree the welfare of any profession is borne by the associations it develops. The officers of the association carry very consequential responsibilities. Their tasks require much time, energy and imagination. It is a continuing but inescapable task to select the gifted members to fill these positions. It is equally important that the officers reflect the long-term interests as well as the wishes of the members of the profession. These matters are part of the larger social problem of making democracy work.

CHAPTER 3

REGULATION OF THE PROFESSION

This chapter deals with the regulation of the profession in Ontario under the Professional Engineers Act, 1984 and Ontario Regulation 538/84 made under that Act. The appropriate Sections and Subsections of the Ontario Act and Regulation are referred to in the text.

For a jurisdiction other than Ontario an applicant for membership or a licence to practise would be concerned with the rules and regulations in that province or territory.

In Canada, regulation of engineers and other professionals is the responsibility of the governments of the provinces and territories. In Ontario, the first Professional Engineers Act was that of 1922; there have been amendments and superseding Acts over the years. The current Act is known by the short title: The Professional Engineers Act, 1984.

At the risk of stating what is obvious to the reader, it is pointed out that the word "Regulation" in the title of this chapter is a noun, meaning "the process of regulation, i.e., of prescribing rules and exercising authoritative direction," although it is common to use the same noun (depending on the context) to designate a particular rule. Also, in the title of this chapter, we intend "the profession" to mean the profession of engineering, as distinct from the general meaning in our discussion of professions in Chapter 2.

Self-Regulation of the Engineering Profession in Canada

If regulation of the profession is, as we say, the responsibility of governments, how can we say, as we also do, that the profession in Canada is self-regulating?

The Acts stipulate who shall, and who shall not, practise professional engineering.

Even a rapid survey of the Acts (of the provinces and territories) will explain this seeming anomaly, and the Ontario Act will serve as illustration. In the first place, the Act provides the constitution of a professional body [The Association of Professional Engineers of the Province of Ontario—see Sections 2(1) and 2(4)]; every person who holds the unrestricted licence of this body to practise professional engineering, as defined in Section 1(m), is a member of it [see Section 5(1)], and no person shall engage in such practice unless he holds an unrestricted licence and is therefore a member of this body, or else holds its temporary licence or limited licence [see Section 12(1)].

The Association is bound to issue a licence to a person who meets the qualifications of Section 14(1) of the Act. Note that academic and experience requirements included in these qualifications [Section 14(1), (c) and (d)] are set, not by the Act, but by the regulations.

Now return to Section 7, under which, subject to the approval of the Lieutenant Governor in Council and with prior review by the Minister, the Association makes the regulations that govern, not only the educational and experience qualifications for membership already mentioned, but all matters of the practice of professional engineering not explicitly covered by the Act. See Regulation 538/84 (known as O. Reg.538/84).

This is the mechanism of self-regulation.

Member and Licensee; Temporary and Limited Licences

In Ontario, since the coming into force of the current Act, September 1, 1984, an individual practitioner must be a holder of a licence, of a temporary licence or of a limited licence; if he holds the first of these he is a member of the Association, which gives him certain additional rights in its internal organization and administration. Prior to the current Act, a member of the Association did not require a licence; the term was used only to designate a permission to a non-member, suitably qualified, but not a resident of Ontario, to carry out some specified professional work within the jurisdiction. Such a permission today would be known as a Temporary Licence (refer O. Reg.538/84, Sections 43, 44 and 45). A Limited Licence (O. Reg. 538/84, Sections 46 and 47) permits some specified practice within the scope of professional engineering to a person of qualifications not adequate for a licence or a temporary licence.

Since the word "member" has the general meaning "constituent part of a complex structure" and the word "licence" normally means "a formal permission to do some specified thing," "licensee" seems a more explicit term than "member" for one who holds a formal permission to practise engineering.

Branch of Registration; Practice in Another Branch

In Ontario every licensee is registered as being in one of the following branches of engineering; normally the branch is determined by the preponderant discipline of his education. The branches are: Chemical and Metallurgical; Civil; Electrical; Mechanical Aeronautical and Industrial; Mining.

A licensee's right to practise is not restricted to his branch of registration. However, when he undertakes work he is not competent to perform by virtue of his training and experience, he is guilty of professional misconduct according to O. Reg. 538/84, Section 86.(2)(h).

Work of Technical Personnel

In Ontario, the Act does not prevent a person who is not a licensee from doing an act that is within the practice of professional engineering where a professional engineer assumes responsibility for the service within the practice of professional engineering to which the act is related. [P.E. Act, 1984, Section 12(3)(b)]

Partnerships and Corporations; the Certificate of Authorization

A partnership or corporation can act only as the persons it comprises, or the managers they elect, dictate. If, therefore, it is to practise professional engineering, and thereby undertake the responsibilities, both legal and ethical, that are imposed upon an individual practitioner (in Ontario, by the Professional Engineers Act, 1984, and Regulation 538/84) and because failure to meet these responsibilities is, for the most part, and particularly in matters of ethics, the failure of an individual, there must be at least one person of adequate authority within the partnership or corporation who, being a licensee of the Association (member) or holder of its temporary licence, and himself or herself the holder of a Certificate of Authorization (see section of this Chapter immediately following) formally undertakes responsibility for, and supervision of, the practice of professional engineering by the corporation or partnership.

This formal undertaking is part of the form of application by which a Certificate of Authorization is requested of the Association. The Certificate of Authorization is a formal permission for the partnership or corporation to practise professional engineering in its own name to the extent of the right to practise of the licensees (members), or holders of temporary licences, who have signed the undertaking (P.E. Act, 1984, Section 15 and O. Reg. 538/84, Sections 48, 49).

Individual Practitioners; the Certificate of Authorization

Retaining its purpose, expressed above in connection with partnerships and corporations, a Certificate of Authorization is now (in Ontario, since the coming into force of the P.E. Act, 1984), a requirement for every individual licensee (member) or holder of a temporary licence where it is or will be the primary function of the individual ". . . to provide to the public services that are within the practice of professional engineering." In a Certificate of Authorization granted to an individual member or temporary licensee, the individual is his or her own guarantor.

Professional Liability Insurance; the Certificate of Authorization

A reason for the requirement (in Ontario, since the coming into force of the P.E. Act, 1984 and O. Reg. 538/84) of a Certificate of Authorization for certain individual practitioners, as set forth in the preceding paragraph, may be that the Certificate of Authorization is to be issued only where the recipient has shown proof of carrying specified minimal professional liability insurance, as stipulated in O. Reg. 538/84, Section 88. This section does not appear in the selected excerpts given in this chapter, because, at the time of going to press, the section is in abeyance.

Reserved Title for Members and Temporary Licensees

The title for members is "Professional Engineer" or abbreviation, usually "P.Eng.", or, in the jurisdiction of Quebec, "ingénieur" or "ing."

Additional Reserved Title: "Consulting Engineer"

In Ontario, members suitably qualified may be so designated by the Council of APEO; this does not extend the member's right to practise, but gives him or her the right to use the additional reserved title. The qualifications are set forth in O. Reg. 538/84, Sections 70, 73, 74.

Power of Control of Licensees

A licensing body, by issuing a licence, declares implicitly that the licensee is competent to engage in the practice the licence permits. In undertaking the responsibility for such a declaration, it must have some power of control. This power is usually represented by the right to impose penalties for demonstrated professional misconduct or for demonstrated incompetence, and all licensees are subject to this power. Penalties may be reprimand, suspension or termination of licence, or the exaction of a fine [as, for example, in Ontario, P.E. Act, 1984, Section 29(4)(h)]. A licensing body has no power to punish any person other than its licensees.

Ontario—The Discipline Committee of the APEO

In the Ontario jurisdiction, again taken as an example, the Discipline Committee "shall . . . hear and determine allegations of professional misconduct or incompetence . . ." . The member or holder of a Certificate of Authorization, a temporary licence or a limited licence, against whom the allegation is made, may be found guilty of professional misconduct if:

- he or she has been found guilty (in a civil court) of an offence relevant to suitability to practise, upon proof of such conviction;

- he or she has been guilty in the opinion of the Discipline Committee of professional misconduct as defined in the regulations,

or may be found incompetent because of:

- lack of knowledge, skill or judgment or disregard for the welfare of the public,
- a physical or mental condition . . . making it desirable that he or she no longer be permitted to practise, or that his or her practice be restricted.

Note: See Ontario Act, Section 29 and O. Reg. 538/84, Section 86.

Professional Misconduct and the Code of Ethics

As will be seen in the excerpts, O. Reg. 538/84 effects a separation of punishable actions of professional misconduct (Section 86) from canons of moral obligation (The Code of Ethics, Section 91). Although, in Section 86, it says that an action that is "solely a breach of the code of ethics" does not constitute an offence of professional misconduct, it has been argued that Section 86(2)(j) might, in an extreme case, negate this exclusion.

The separation spoken of above is analogous to practice in the criminal courts: if I am charged with a wilful act causing damage to another person, the charge is one of libel, attempted murder, or whatever, but there is no charge against me for breaking the moral law expressed (in the Judaeo-Christian tradition) as "Thou shalt love thy neighbour as thyself." There is no action of professional misconduct that does not breach the Code of Ethics in one way or another, but it is for the action of professional misconduct that a practitioner is answerable to the Association.

Two Sections of O. Reg. 538/84, Section 86, Professional Misconduct and Section 91, the Code of Ethics are included in their entirety in the Appendix of this book.

The Code of Ethics

Consider a case where there is a breach of the Code of Ethics, but no act of professional misconduct. Here the practitioner must sit in judgment on himself. Take, as an imaginable example, a practitioner who, as designer of what the newspapers call the "soaring structure" of a major civic project, is the principal speaker at a banquet celebrating completion. His speech is deft, modest, pleasing to the assembled worthies, and is widely quoted. He is generous in his political credits, but fails to mention the collaborating engineering firm responsible for the unusually difficult foundation design. Within the profession, the practitioner is perceived as having breached s. 91.7 i of the Code of Ethics (discourtesy towards another practitioner), and some ask themselves whether there is not a more serious breach, a covert and malicious attempt to injure the reputation and business of a fellow practitioner (s. 91.7 iii). Only the conscience of the offender can reveal the motives, and it is in the motives that the degree of moral guilt exists, that the character must be judged as being—or not being—that which is unequivocally described in s. 1 of the Code of Ethics. Self-judgment, certainly—yet it is not so entirely private a matter. Justly or not, the conclusion may be drawn that a malicious or vindictive motive has been present, and the practitioner has no way of knowing what is being thought about him, nor will he have a hearing before his peers at which he can defend himself. In some way, a minor ethical breach, not itself punishable, may ill affect the career of a practitioner. As we contemplate, alone under the midnight oil, how we will behave in a given situation, what we will write, what we will say at tomorrow's meeting, what is good faith, what is adequate knowledge and conviction, aware of our liberty to interpret the Code of Ethics, it will be well that we bear in mind that practical consequences may exist, as imagined in our example, and remember this sharp-edged little Spanish proverb:

> *Take what you will, says God—and pay for it.*

Consequences more immediate and more material than those imagined in the foregoing example may follow an ethical decision a practitioner has to face. The few inspiring words of Section 1 of the Code of Ethics tell us what we should be rather than what we should do. This advice contains the elements of a decision most of us hope never to have to make—a decision arising from a perceived conflict between loyalty to an employer and fidelity to public needs. What is the engineer's duty in the matter when the practices or the plans of an employer which adversely affect the public welfare are unlikely to become known to the public, or understood by it, until too late?

To illustrate this, let us consider four situations, each of which requires consideration and a decision by you, the engineer:

(i) Trivial—the employer's plans impose an inconvenience—people will have to walk around, rather than across.

(ii) Real—but countenanced by law with certain restrictions—gaseous or particulate effluent is an example.

(iii) Real—but remote from the engineer's area of responsibility—let us say that the employer plans to incorporate in his detergent product a small amount of a substance that is legally barred, but barred only as a constituent of prepared foods, because it is dangerous if ingested. The presence of the substance is not to be shown on the label; it might scare the buying public.

(iv) Real—the engineer is instructed to design and use a change in the process of manufacture of a product which meets the requirements of the appropriate CSA Standard. The appearance of the product would not be affected. The engineer, who thoroughly understands the process, is convinced that the change, which is essentially a shortcut, will reduce the life of the product and introduce a new risk of accident to the user.

In the first example, we have called the matter "trivial," referring to an "inconvenience" inflicted on the public, intending to portray a small matter in which no reasonable ethical worry exists for the engineer. However, it makes one point; the harm from which it is the engineer's ethical duty to protect the public must have some real substance; interpretation of the relevant canon of the Code of Ethics requires judgment as to degree of harm or need that appears; our ethical duty does not require us to set a lance in rest and tilt at trifles.

The second case deals with a continuing practice of the employer that is likely to have worse consequences for the public than a mere inconvenience; nevertheless, provided the legal restrictions (presumably rating the degree of harm as tolerable) are faithfully met, the engineer's ethical duty seems to be fulfilled by his constantly studying the process and the progress of smoke abatement technology, and by urging his employer to use such improvements as may be economically possible—economically, because it is also in the interest of the public that the plant should thrive and continue to provide work. Anyone who has lived in a newsprint paper mill town, where sulphite pulp digesters are blown periodically to the atmosphere, will realize that public interest is itself a complex matter.

In case (iii) the engineer, worried that there might be some failure in his own duty to the public wellbeing regarding his employer's intention to spike the detergent with a substance that has been found harmful in another use, would have to analyze his position in the matter, an undertaking we cannot simulate without knowing a range of circumstances not given in the question. We can imagine extremes: the principal factor is the engineer's knowledge of biochemistry and whether it is sufficient to warrant a substantial doubt of the safety of the additive in the proposed use (implying also a doubt of the comprehensiveness of the Health Department's research). If he decides that his knowledge is sufficient, and protests vigorously but unsuccessfully to his employer, he might have to go farther, even to the bitter end of "blowing the whistle," an outcome imagined in the fourth of these hypothetical situations. The other extreme is where the engineer has no knowledge of the subject as a basis for his suspicion; the very name of the additive has scared him, just as the employer figures it would scare the consumer.

In the fourth example, the engineer's ethical duty to protest is clear (Sections 91.1 and 91.2. i of the Code of Ethics). He does so, face-to-face with the plant manager (taken here to represent the employer, probably a corporation). The manager at first makes light of the engineer's warning. "It's the same end product—just a change in our way of making it," he says. "It's the way the competition does it, and we can't afford costs higher than theirs . . . ." And so forth. "Sure, I know you've got the experience and a first class record, but this is the way it's got to be." Before long the talk is fraught with anger; it ends in harsh words: "I won't go along with it!"

"Okay. Then, dammit, we'll get someone who will."

So what happens? Tempers cool. The two avoid each other. Both spend a few restless nights. It is the manager who makes the next move; he summons a closed door conference of half a dozen senior and confidential employees of the company, a group that includes the engineer, who explains his objections in detail, using a blackboard to illustrate his contention because not all present are familiar with the construction of the product—indeed, it is the old chief accountant who really defines the issue: "This," he says, "is the sort of thing Ruskin had in mind when he wrote—how does it go?—'There is nothing in this world that some man cannot make a little worse and sell a little cheaper.' If you and our engineer are both right, the proposal is that we join our competitors by putting out a worse article dressed in the sheep's clothing of the C.S.A. Standard."

The tenor of the argument is now changed. Again, what happens? Is the engineer's opinion finally accepted, or a compromise satisfactory to him reached, such as, say, a decision to install additional equipment that will eliminate the adverse effects of the process change? Or does it go the other way, for endings are not always happy? Suppose it does. The engineer, clear in his conscience of any motive of vindictiveness or of personal gain, sadly packs up his gear and "blows the whistle"—in this case probably by warning the authority having jurisdiction of the impending change and of his opinion as to the risk it imposes on the public. Personal gain—damages for dismissal, a court order for reinstatement? The first is likely to be small recompense for the hardship and the heartache he has undergone; of the latter he will not take advantage. There is no equivalent employment in the immediate offing. Yet he may well have gained more than he has lost. For the engineering profession distinguishes between the men and the boys among its members; reputation is an asset that increases in value as the years of practice progress, and there is no aspect of reputation greater than that of incorruptible personal honour.

We go back to the Spanish proverb already cited as applying to the lonely business of reaching an ethical decision, and re-write it with an addendum:

> *Take what you will, says God—and pay for it.*
> *And know that there is a credit side to the account!*

Professional Misconduct

We quote here verbatim the content of s. 86, O.Reg. 538/84, Professional Misconduct, interjecting remarks which are intended to illustrate some of the clauses by practical examples or to justify by the authority of the clause opinions we have expressed elsewhere in this book. Our interpretations have no official standing.

> (1) In this section, "negligence" means an act or an omission in the carrying out of the work of a practitioner that constitutes a failure to maintain the standards that a reasonable and prudent practitioner would maintain in the circumstances.

". . . the standards that a reasonable and prudent practitioner would maintain in the circumstances" seems to express the level of care considered as normal in legal actions. Marston, in his *Law for Professional Engineers*, 2nd ed., pp. 28 to 30, under the heading "The Engineer's Standard of Care" cites instances and opinions from the courts to illustrate interpretations of what is "reasonable."

> (2) For the purposes of the Act and this Regulation, "professional misconduct" means,
> (a) negligence;
> (b) failure to make reasonable provision for the safeguarding of life, health or property of a person who may be affected by the work for which the practitioner is responsible;

The act or omission that constitutes the failure in (b) above is explicitly an offence of professional misconduct with which a practitioner may be charged; he is not charged with a breach of the Code of Ethics, although, certainly, he has disregarded his conscience and ignored his ethical duty.

> (c) failure to act to correct or report a situation that the practitioner believes may endanger the safety or the welfare of the public;

What is said about (b) applies also to (c). To "report a situation," if one cannot correct it, is commonly known as "whistle blowing," a modern expression for the ultimate recourse of the practitioner whose conscience forbids him to act, or to condone an act of his employer or client, where, in his carefully considered opinion, a result harmful to the public interest will ensue. Earlier in this chapter, under the subhead "The Code of Ethics," we discussed (in case iv) an imaginary instance of such a situation, treating it as an ethical problem for the practitioner. In fact, since he is represented as believing the risk to the public to be real, and since he has been unable to correct it, he is guilty of professional misconduct under this clause 86(2)(c) if he does not "blow the whistle."

> (d) failure to make responsible provision for complying with applicable statutes, regulations, codes, by-laws and rules in connection with work undertaken by or under the responsibility of the practitioner;

Knowledge of such constraints upon design and execution of the work is properly expected by an employer or client. Although a disciplinary charge under this subsection appears to be not far from a charge of negligence, the constraints contemplated here seem to be civic rather than professional.

> (e) signing or sealing a drawing, specification, plan, report or other document not actually prepared or checked by the practitioner;

The very act of providing a signature or seal and thereby guaranteeing the responsibility of the practitioner where the practitioner is not fully cognizant of the contents of drawing or document is the offence; the content does not have to be faulty in order that a disciplinary action may be instituted under this subsection.

> (f) failure of a practitioner to present clearly to his employer the consequences to be expected from a deviation proposed in work, if the professional engineering judgment of the practitioner is overruled by non-technical authority in cases where the practitioner is responsible for the technical adequacy of professional engineering work;

A practitioner's duty to his employer certainly requires that he inform that employer if deviations in the technical plans have been made against his advice and that he explain the ill consequences he foresees. If the deviation is carried out in spite of this warning, the harm will be to the employer. It would be the part of wisdom for the practitioner to express his warning in writing; provided the public interest is not affected, he need go no farther.

> (g) breach of the act or regulations, other than an action that is solely a breach of the code of ethics;

A practitioner breaching the Act is probably doing so as a member of the public, since the Act's prohibitions apply almost entirely to "any person"; in his professional affairs the practitioner is normally subject to the governance of the Regulation. The exclusion of a breach of the Code of Ethics as professional misconduct expressed in this subsection might be overridden in an extreme case, as represented by subsection (j).

(h) undertaking work the practitioner is not competent to perform by virtue of his training and experience;
(i) failure to make prompt, voluntary and complete disclosure of an interest, direct or indirect, that might in any way be, or be construed as, prejudicial to the professional judgment of the practitioner in rendering service to the public, to an employer or to a client, and in particular without limiting the generality of the foregoing, carrying out any of the following acts without making such a prior disclosure;

1. Accepting compensation in any form for a particular service from more than one party.
2. Submitting a tender or acting as a contractor in respect of work upon which the practitioner may be performing as a professional engineer.
3. Participating in the supply of material or equipment to be used by the employer or client of the practitioner.
4. Contracting in the practitioner's own right to perform professional engineering services for other than the practitioner's employer.
5. Expressing opinions or making statements concerning matters within the practice of professional engineering of public interest where the opinions or statements are inspired or paid for by other interests.

Discussing severally the offences 1 through 5 above as typical of a whole range of imaginable situations [i.e., typical of the generality of what is enjoined in subsection (i)] we find ourselves in the complex area of conflict of interest, complex principally because of the ethical obligations that attend the simple act of disclosure to employer or client whereby the innocence of the practitioner appears to be protected.

1. If a practitioner accepts compensation for the same work from two or more clients or employers and each of these is aware of and has agreed to the participation of the others, there must certainly be no adversarial elements, at this stage, in the relations between the several clients or employers. So long as each of these receives the complete rendering of the engineer's findings, this situation seems to be no more than a cooperative funding of an engineering work. The presence of a secret participant in the practitioner's compensation would change the whole to a grave matter of profes-

sional misconduct; indeed, it is difficult to imagine any innocent motive that could be brought as a plea in defence against such a charge.

Even a gift to the practitioner by a party whose interest may be adverse to that of the practitioner's employer or client—the bottle of 12-year-old Scotch at Christmas, say, representing some previous association or present friendship—will be acknowledged by the careful practitioner in writing with a copy to his employer or client.

2. & 3. A practitioner who prepares a design or specification to go out for tender, and has an interest in any of the companies that will bid on it, has fulfilled the letter of subsection (i) if, from the start of the activity, he has completely divulged this interest to his employer or client. The ethical obligation that devolves upon him is that the design or specification be not deliberately drawn so as to render the conditions to be met by the contractor impossible, or extraordinarily difficult, for any but the company in which the practitioner is interested, so that the calling for tenders is essentially a fake, and the engineer is acting as a covert salesman.

There can be innocent versions of such a situation: the law may require that the work go out for tender, but there may be reasons satisfactory to the employer or client why only one source of supply can fulfil the requirements, and the specification names that source. Matching of existing equipment, apparatus of high and closely held technology, and so on, for a government-owned organization might be examples of this. A call for tenders cannot be labelled a fake if it is written so that it is quite clear that it cannot be a matter of competition, at least as to the origin of the principal equipment or expertise to be supplied.

4. A practitioner employed full-time who concurrently performs engineering services for one or more additional employers or clients is in an anomalous position as compared with that visualized in example 1 above: certainly his prime duty is to the full-time employer, and a relationship of mutual trust and confidentiality must be presumed to exist between them. If additional employers or clients are in the least degree in competition with the principal employer it is hard to imagine the granting of the necessary permission by the latter which must have accompanied the "prompt, voluntary and complete disclosure" made to him by the practitioner.

A notable example of complete innocence in such a situation is that of the academic practitioner whose practical knowledge and experience are kept up to date by concurrent professional engineering work, undertaken with the full knowledge and approval of the college or university that is his full-time employer.

5. A case in point is that of the expert witness, who is permitted to express professional opinions where non-expert witnesses are normally restricted to facts. His services have usually been enlisted by one of the disputants, whose counsel has previously ascertained that his opinion tends to favour its case. The most rigorous ethical sense must govern his statements; that he has an interest (his fee for the service) is already publicly divulged, and the requirement of subsection (1) thereby met; now his conscience must govern his tongue so that no consideration other than his unbiassed professional judgment enters his evidence.

A practitioner who offers a statement for publication on professional engineering aspects of some project of public interest which has not yet been committed for engineering study to another practitioner, is exercising the right of a citizen. It would be the part of wisdom for him to include a statement that he has no private knowledge of or influence in the matter. If such an addendum cannot be truthfully made, he is at risk with respect to subsection (i) and to his ethical self-respect.

> (j) conduct or an act relevant to the practice of professional engineering that, having regard to all the circumstances, would reasonably be regarded by the engineering profession as disgraceful, dishonourable or unprofessional.

The Act itself [s. 29(2)(a)] renders an offence "relevant to suitability to practise" of which a practitioner has been convicted (in the courts) grounds for a charge of professional misconduct before the Discipline Committee. Under this subsection (j) of the Regulation, if the conduct or act described is "relevant to the practice of professional engineering," no such conviction is a prerequisite to a charge of professional misconduct.

> (k) failure by a practitioner to abide by the terms, conditions or limitations of the practitioner's licence, limited licence, temporary licence or certificate;
> (l) failure to supply documents or information requested by an investigator acting under section 34 of the Act;
> (m) permitting, counselling or assisting a person who is not a practitioner to engage in the practice of professional engineer-

ing except as provided for in the Act or the regulations O. Reg. 538/84, s. 86.

The three subsections above are among the few exceptions to our several statements in this book that it is normally the Regulation rather than the Act that governs the conduct of the individual practitioner; the offences (k), (l) and (m) are illegal as well as being professional misconduct. With respect to (m), see, earlier in this chapter, what is said under the heading "Professional and Technical Personnel."

Ontario—The Professional Engineers Act, 1984; Selected Excerpts

The Act is officially described as "Professional Engineers Act, Bill 123, Chapter 13, Statutes of Ontario." Your authors gratefully acknowledge permission of the Association to reproduce these excerpts from its Office Consolidation edition, of which copies are available at the APEO office. Sections of both Act and Regulation cited in this chapter are included in the excerpts chosen.

HER MAJESTY, by and with the advice and consent of the Legislative Assembly of the Province of Ontario, enacts as follows:

1. In this Act,

(m) "practice of professional engineering" means any act of designing, composing, evaluating, advising, reporting, directing or supervising wherein the safeguarding of life, health, property or the public welfare is concerned and that requires the application of engineering principles, but does not include practising as a natural scientist;

(n) "professional engineer" means a person who holds a licence or a temporary licence;

(o) "Registrar" means Registrar of the Association;

(p) "regulations" means regulations made under this Act;

(q) "temporary licence" means temporary licence to engage in the practice of professional engineering issued under this Act.

2.—(1) The Association of Professional Engineers of the Province of Ontario, a body corporate, is continued as a corporation without share capital under the name of "Association of Professional Engineers of Ontario".

(4) For the purpose of carrying out its principal object, the Association has the following additional objects:

1. To establish, maintain and develop standards of knowledge and skill among its members.
2. To establish, maintain and develop standards of qualification and standards of practice for the practice of professional engineering.
3. To establish, maintain and develop standards of professional ethics among its members.
4. To promote public awareness of the role of the Association.
5. To perform such other duties and exercise such other powers as are imposed or conferred on the Association by or under any Act.

5.—(1) Every person who holds a licence is a member of the Association subject to any term, condition or limitation to which the licence is subject.

7.—(1) Subject to the approval of the Lieutenant Governor in Council and with prior review by the Minister, the Council may make regulations.

9. respecting any matter ancillary to the provisions of this Act with regard to the issuing, suspension and revocation of licences, certificates of authorization, temporary licences and limited licences, including but not limited to regulations respecting,
   i. the scope, standards and conduct of any examination set or approved by the Council as a licensing requirement,
   ii. the curricula and standards of professional training programs offered by the Council,
   iii. the academic, experience and other requirements for admission into professional training programs,
   iv. classes of licences,
   v. the academic and experience requirements for the issuance of a licence or any class of licence, and
   vi. classes of certificates of authorization, temporary licences and limited licences, including prescribing requirements and qualifications for the issuance of specified classes of certificates of authorization, temporary licences and limited licences, and terms and conditions that shall apply to specified classes of certificates of authorization, temporary licences and limited licences;
10. prescribing forms of applications for licences, certificates of authorization, temporary licences and limited licences and requiring their use;

32. specifying acts within the practice of professional engineering that are exempt from the application of this Act when performed or provided by a member of a prescribed class of persons, and prescribing classes of persons for the purpose of the exemption.

11. The Council may delegate to the Executive Committee the authority to exercise any power or perform any duty of the Council other than to make, amend or revoke a regulation or a by-law.

12.—(1) No person shall engage in the practice of professional engineering or hold himself out as engaging in the practice of professional engineering unless the person is the holder of a licence, a temporary licence or a limited licence.

(2) No person shall offer to the public or engage in the business of providing to the public services that are within the practice of professional engineering except under and in accordance with a certificate of authorization.

(3) Subsections (1) and (2) do not apply to prevent a person,

(a) from doing an act that is within the practice of professional engineering in relation to machinery or equipment, other than equipment of a structural nature, for use in the facilities of the person's employer in the production of products by the person's employer;

(b) from doing an act that is within the practice of professional engineering where a professional engineer assumes responsibility for the services within the practice of professional engineering to which the act is related;

(c) from designing or providing tools and dies;

(d) from doing an act that is within the practice of professional engineering but that is exempt from the application of this Act when performed or provided by a member of a class of persons prescribed by the regulations for the purpose of the exemption, if the person is a member of the class;

(e) from doing an act that is exempt by the regulations from the application of this Act.

14.—(1) The Registrar shall issue a licence to a natural person who applies therefor in accordance with the regulations and,

(a) is a citizen of Canada or has the status of a permanent resident of Canada;

(b) is not less than eighteen years of age;

(c) has complied with the academic requirements specified in the regulations for the issuance of the licence and has passed such examinations as the Council has set or approved in accordance with the regulations or is exempted therefrom by the Council;

(d) has complied with the experience requirements specified in the regulations for the issuance of the licence; and

(e) is of good character.

15.—(1) The Registrar shall issue a certificate of authorization to a natural person, a partnership or a corporation that applies therefor in accordance with the regulations if the requirements and qualifications for the issuance of the certificate of authorization set out in the regulations are met.

17.—(1) It is a condition of every certificate of authorization that the holder of the certificate shall provide services that are within the practice of professional engineering only under the personal supervision and direction of a member of the Association or the holder of a temporary licence.

(2) A member of the Association or a holder of a temporary licence who personally supervises and directs the providing of services within the practice of professional engineering by a holder of a certificate of authorization or who assumes responsibility for and supervises the practice of professional engineering related to the providing of services by a holder of a certificate of authorization is subject to the same standards of professional conduct and competence in respect of the services and the related practice of professional engineering as if the services were provided or the practice of professional engineering was engaged in by the member of the Association or the holder of the temporary licence.

20. A corporation that holds a certificate of authorization has the same rights and is subject to the same obligations in respect of fiduciary, confidential and ethical relationships with each client of the corporation that exist at law between a member of the Association and his client.

29.—(1) The Discipline Committee shall,

(a) when so directed by the Council, the Executive Committee or the Complaints Committee, hear and determine allegations of professional misconduct or incompetence against a member of the Association or a holder of a certificate of authorization, a temporary licence or a limited licence;

(2) A member of the Association or a holder of a certificate of authorization, a temporary licence or a limited licence may be found guilty of professional misconduct by the Committee if,

(a) the member or holder has been found guilty of an offence relevant to suitability to practise, upon proof of such conviction;

(b) the member or holder has been guilty in the opinion of the Discipline Committee of professional misconduct as defined in the regulations.

(3) The Discipline Committee may find a member of the Association or a holder of a temporary licence or a limited licence to be incompetent if in its opinion,

(a) the member or holder has displayed in his professional responsibilities a lack of knowledge, skill or judgment or disregard for the welfare of the public of a nature or to an extent that demonstrates the member or holder is unfit to carry out the responsibilities of a professional engineer; or

(b) the member or holder is suffering from a physical or mental condition or disorder of a nature and extent making it desirable in the interests of the public or the member or holder that the member or holder no longer be permitted to engage in the practice of professional engineering or that his practice of professional engineering be restricted.

(4) Where the Discipline Committee finds a member of the Association or a holder of a certificate of authorization, a temporary licence or a limited licence guilty of professional misconduct or to be incompetent it may, by order,

(a) revoke the licence of the member or the certificate of authorization, temporary licence or limited licence of the holder;

(b) suspend the licence of the member or the certificate of authorization, temporary licence or limited licence of the holder for a stated period, not exceeding twenty-four months;

(c) accept the undertaking of the member or holder to limit the professional work of the member or holder in the practice of professional engineering to the extent specified in the undertaking;

(d) impose terms, conditions or limitations on the licence or certificate of authorization, temporary licence or limited licence, of the member or holder, including but not limited to the successful completion of a particular course or courses of study, as are specified by the Discipline Committee;

(e) impose specific restrictions on the licence or certificate of authorization, temporary licence or limited licence, including but not limited to,

(i) requiring the member or the holder of the certificate of authorization, temporary licence or limited licence to engage in the practice of professional engineering only under the personal supervision and direction of a member.

(ii) requiring the member to not alone engage in the practice of professional engineering.

(iii) requiring the member or the holder of the certificate of authorization, temporary licence or limited licence to accept periodic inspections by the Committee or its delegate of documents and records in the possession or under the control of the member or the holder in connection with the practice of professional engineering,

(iv) requiring the member or the holder of the certificate of authorization, temporary licence or limited licence to report to the Registrar or to such committee of the Council as the Discipline Committee may specify on such matters in respect of the member's or holder's practice for such period of time, at such times and in such form, as the Discipline Committee may specify;

(f) require that the member or the holder of the certificate of authorization, temporary licence or limited licence be reprimanded, admonished or counselled and, if considered warranted, direct that the fact of the reprimand, admonishment or counselling be recorded on the register for a stated or unlimited period of time;

(g) revoke or suspend for a stated period of time the designation of the member or holder by the Association as a specialist, consulting engineer or otherwise;

(h) impose such fine as the Discipline Committee considers appropriate, to a maximum of $5,000, to be paid by the member of the Association or the holder of the certificate of authorization, temporary licence or limited licence to the Treasurer of Ontario for payment into the Consolidated Revenue Fund;

(i) subject to subsection (5) in respect of orders of revocation or suspension, direct that the finding and the order of the Discipline Committee be published in detail or in summary and either with or without including the name of the member or holder in the official publication of the Association and in such other manner or medium as the Discipline Committee considers appropriate in the particular case;

(j) fix and impose costs to be paid by the member or the holder to the Association;

(k) direct that the imposition of a penalty be suspended or postponed for such period and upon such terms or for such purpose, including but not limited to,

(i) the successful completion by the member or the holder of the temporary licence or the limited licence of a particular course or courses of study.

(ii) the production to the Discipline Committee of evidence satisfactory to it that any physical or mental handicap in respect of which the penalty was imposed has been overcome,

or any combination of them.

35. It is a condition of every certificate of authorization that the holder of the certificate shall not offer or provide to the public services that are within the practice of professional engineering unless the holder is insured in respect of professional liability in accordance with the regulations.

51.—(1) The *Professional Engineers Act*, being chapter 394 of the Revised Statutes of Ontario, 1980, is repealed.

53. The short title of this Act is the *Professional Engineers Act, 1984.*

Ontario—Regulation 538/84; Selected Excerpts

The full title of this document is "Regulation 538/84, made under the Professional Engineers Act (Chapter 13, Statutes of Ontario, 1984)." As with the excerpts from the Act immediately preceding, your authors are indebted to the Association for permission to reproduce these from the Regulation, using the Office Consolidation edition, of which copies are available at the APEO office. Again, the selection is made so as to include all sections cited in this chapter.

Applicants for Licence, Education and Experience

34.(1) Each applicant for a licence shall,

(a) demonstrate,

i. that he has earned a bachelor's degree in an engineering program from a Canadian university that is accredited to the satisfaction of the Council, or

ii. that he has equivalent engineering educational qualifications recognized by the Council;

(b) demonstrate twenty-four months of experience following the conferring of a degree or the completion of equivalent engineering education, as the case may be, in the practice of professional engineering that will provide sufficient experience to enable the applicant to meet the generally accepted standards of practical skill required to engage in the practice of professional engineering; and

(c) successfully complete the Professional Practice Examination.

(2) Twelve months of the experience mentioned in clause (1)(b) must be experience in Canada under the supervision of one or more persons legally authorized to engage in the practice of professional engineering in the jurisdiction in which the experience was acquired.

(3) The Council, in circumstances where it considers it in the public interest to do so, may vary or waive the requirement in subsection (2) as to twelve months of experience in Canada. O. Reg. 538/84, s. 34.

Examinations

35. Examinations required by the Academic Requirements Committee shall be held prior to the 1st day of June in each year and at such other time, if any, and at such place or places, as the Council may from time to time determine. O.Reg. 538/84, s. 35.

38. An applicant for a licence must pass the Professional Practice Examination not later than two years following the later of,

(a) the date of submission of the application for membership by the applicant to the Registrar; and

(b) the date of successful completion of all other examination requirements (other than the writing of the thesis, if required) or the final determination that no examination or thesis is required. O. Reg. 538/84, s. 38.

Experience Requirements Committee

42.(1) There is hereby established the Experience Requirements Committee composed of a chairman appointed by Council, the immediate past chairman, if any, and such other Members as are appointed by the Council, and three members of the Committee constitute a quorum.

(2) Where an application for the issuance of a licence, temporary licence or limited licence is referred to the Experience Requirements Committee pursuant to the Act, the Committee shall,

(a) assess the experience qualifications of the applicant; and

(b) determine whether the applicant meets the experience requirements prescribed by this Regulation and so advise the Registrar.

(3) For the purpose of carrying out its duties, the Experience Requirements Committee may, in the discretion of the Committee and on its own initiative, interview the applicant.

(4) The Committee shall interview the applicant if there is a question raised with respect to the ability of the applicant to communicate adequately in the English language. O. Reg. 538/84, s. 42.

Temporary Licences

43.(1) Every temporary licence must specify,

(a) the works, facilities, machinery, equipment or other property in Ontario to which the temporary licence relates;

(b) the name of the person, firm or corporation by whom the holder of the temporary licence is employed or engaged to perform services in Ontario within the practice of professional engineering;

(c) the name of the Member, if any, with whom collaboration is required under this Regulation; and

(d) the period of time, not exceeding twelve months, for which the temporary licence has been issued.

(2) It is a condition of every temporary licence that the services within the practice of professional engineering that may be provided by the holder of the temporary licence are limited to the services specified in the temporary licence. O. Reg. 538/84, s. 43.

44. The requirements and qualifications for the issuance of a temporary licence are payment of the fee for the temporary licence and one of the following:

1. Residence in a province or territory of Canada other than Ontario and membership in an association of professional engineers in another province or territory of Canada that has objects similar to the objects of the Association and that requires qualifications for membership at least equal to the qualifications required for the issuance of a licence to engage in the practice of professional engineering in Ontario.

2. Qualifications at least equal to the qualifications required for the issuance of a licence to engage in the practice of professional engineering in Ontario.
3. Wide recognition in the field of the practice of professional engineering in respect of which the work to be undertaken under the temporary licence relates and not less than ten years experience in such field. O. Reg. 538/84, s. 44.

45.(1) It is a term and condition of every temporary licence that the holder of the temporary licence must collaborate with a Member in the practice of professional engineering in respect of the work undertaken under the temporary licence unless the holder,

(a) is a member of an association of professional engineers in another province or territory of Canada that has objects similar to the objects of the Association and that requires qualifications for membership at least equal to the qualifications required for the issuance of a licence under this Act;

(b) provides evidence that he has qualifications at least equal to the qualifications required for the issuance of a licence under this Act and that he is knowledgeable about all codes, standards and laws relevant to the work undertaken under the temporary licence;

(c) provides evidence of wide recognition in the field of the practice of professional engineering relevant to the work undertaken under the temporary licence and that he is knowledgeable about all codes, standards and laws relevant to the work undertaken under the temporary licence; or

(d) is providing services within the practice of professional engineering outside Ontario that are required by another Act to be performed by a professional engineer.

(2) It is a term and condition of every temporary licence held by a person who must collaborate with a Member that the holder of the temporary licence must not issue a final drawing, specification, plan, report or other document unless the Member has signed, dated and affixed his seal thereto. O.Reg. 538/84, s. 45.

Limited Licence

46. The following conditions apply to every limited licence:

1. The practice of professional engineering by the holder of the limited licence must be limited to the services specified in the limited licence.

2. The practice of professional engineering by the holder of the limited licence must be limited to work in the employ of the employer named in the limited licence.
3. When the holder of the limited licence ceases to be employed by the employer named in the limited licence, the holder must notify the Registrar and return to the Registrar the limited licence and the seal issued to the holder. O.Reg. 538/84, s. 46.

47. The requirements and qualifications for the issuance of a limited licence are:

1. One or more of the following:
   i. A three-year diploma in engineering technology or a Bachelor of Technology degree in engineering technology from an institution approved by the Council.
   ii. A four-year honours science degree in a discipline and from a university approved by the Council.
   iii. Academic qualifications accepted by the Council as equivalent to a diploma or degree mentioned in subparagraph i or ii.
2. Thirteen years of experience in engineering work acceptable to the Council, including the years spent in obtaining the post-secondary academic training referred to in paragraph 1 with at least one year of such experience under the supervision and direction of a Member or Members or under the supervision of a person authorized to practice professional engineering in the province or territory in Canada in which the experience was acquired and at least the last two years of the experience in the services within the practice of professional engineering with respect to which the limited licence is to apply.
3. Payment of the fee prescribed by this Regulation for a limited licence.
4. Successful completion of the Professional Practice Examination.
5. Good character.
6. In the case of an applicant for a limited licence who has not previously held a limited licence, at least the last year of the experience referred to in paragraph 2 must have been with the present employer.
7. A holder of a limited licence who returns the limited licence and related seal to the Registrar and who again becomes employed is entitled to be issued a new limited licence and related seal limited to the services specified in the previous limited licence. O. Reg. 538/84, s. 47.

Certificate of Authorization

48. The requirements and qualifications for the issuance of a certificate of authorization are:

1. The applicant must designate one or more Members or holders of temporary licences as the person or persons who will assume responsibility for and supervise the services to be provided by the applicant within the practice of professional engineering.
2. The application for the certificate of authorization must state that the persons named in paragraph 1 are,
   i. the applicant for the certificate of authorization,
   ii. employees of the applicant,
   iii. partners in the applicant, or
   iv. employees of partners in the applicant,
   and will devote sufficient time to the work of the applicant to carry out the responsibilities set out in paragraph 1.
3. The applicant must file proof of insurance in respect of professional liability in accordance with the regulations in the form that shall be supplied by the Registrar. O. Reg. 538/84, s. 48.

49.(1) A natural person, partnership or corporation that desires a certificate of authorization shall submit an application in the form that shall be provided by the Registrar containing,

(a) the names and addresses of the natural person, all partners, or all officers and directors, as the case may be, of the applicant;

(b) the names of the natural person, partners or employees, as the case may be, who hold licences or temporary licences and who will assume responsibility for and supervise the services provided that are within the practice of professional engineering on its behalf;

(c) the certificate of a person named in clause (b) certifying,
   i. that the information contained in the application is true and correct, and
   ii. in the case of an application for a general certificate of authorization, that the primary function of the applicant is or will be to provide services in the practice of professional engineering to the public.

(2) The information listed in subsection (1) shall be noted on the registers maintained by the Registrar.

(3) The Council may publish the information referred to in subsection (2) from time to time. O. Reg. 538/84, s. 49.

Designation of Consulting Engineers

70.(1) The Council shall designate as a consulting engineer every applicant for the designation who,

(a) is a Member,

(b) is currently engaged, and has been continuously engaged, for not less than two years or such lesser period as may be approved by the Council, in the independent practice of professional engineering in the Province of Ontario;

(c) has had five or more years of experience that is satisfactory to the Council and that is in excess of the minimum requirements to become a Member at the time of such application; and

(d) has passed the examinations prescribed by the Council or has been exempted therefrom, pursuant to subsection (2).

(2) The Council may exempt an applicant from any of the examinations mentioned in clause (1)(d) where the Council is of the opinion that the applicant has appropriate qualifications. O. Reg. 538/84, s. 70.

73. A member who has been designated or redesignated as a consulting engineer may use the title "consulting engineer" or a variation thereof approved by Council from time to time so long as the Member is in the independent practice of professional engineering and the designation or redesignation is valid. O. Reg. 538/84, s. 73.

74. For the purpose of this Regulation, a Member shall be deemed to be in the independent practice of professional engineering if,

(a) the Member is engaged primarily in offering services within the practice of professional engineering to the public and holds a certificate of authorization; or

(b) the Member is a member of a partnership that is the holder of a certificate of authorization or is an employee of a holder of a certificate of authorization and is listed on the application of the holder of the certificate of authorization as a Member designated to be responsible for and to supervise the practice of professional engineering on behalf of the holder. O. Reg. 538/84, s. 74.

The Experience Requirement for Registration

An applicant who has completed the working experience requirement of a professional regulating authority may certainly be held to have gained at least something of three tangible benefits: command of practical skills appropriate to an engineer (and for the most part not subjects of instruction in the universities); familiarity with some branch of industry and, in general, with commercial customs; practice of professional engineering under supervision.

In the eighteenth and nineteenth centuries, the novice who had completed a pupilage (apprenticeship) under a practitioner should have acquired the same benefits plus an additional benefit: an engineering education, of which the substance was whatever the practitioner thought desirable, or whatever the practitioner was willing and able to provide. Today's working experience requirement survives from the apprenticeship system. It is a training process completely separate from the academic, the burden of formal education in engineering having been shifted to the schools.[1] It is intended to preserve what was good in the pupilage system—skills, familiarity with industry and commerce, work under practitioners. It is the last of these that concerns us most in discussing professional ethics.

The standards of ethical behaviour of 200 years ago were set by individual practitioners, who were for the most part "ingenious mechanics," as C. R. Young put it.[2] The pupil, therefore, had as an example an application of personal ethics; examples must have varied in quality between good and bad depending on the character of the practitioner to whom the student was articled. For better or for worse in this respect, the contact was prolonged; periods approaching 10 years were not uncommon in these indentures. Today's professional engineers work with conscious observance of a code of ethics, but the required period of contact of the engineer-in-training with the practitioner is comparatively brief. Registering authorities may therefore be expected to scan carefully this aspect of a record of experience submitted for consideration.

Fortunately, the schools provide a valuable association between the student and the teaching staff, most of the members of which are professionally qualified. At a time far enough removed from student days to give us a fair perspective, we recall such associations (happily, for the most part), and we realize from time to time that in the structures of vertical thinking between first principles and consummation of the level attained and in the lateral sowing of speculative thought, we were generally treated as if we were already professional engineers and that the teaching and manners to which we were subjected were intended to, and did, convey more than instruction in technological argument, even to the point of unspoken suggestion of ethics. Nevertheless, in an experience record, it is the presence of a reasonable contact with professional engineers in the working environment rather than in the academic environment that is normally demanded.

There used to be certain skills expected of the graduate engineer, presumably to be further developed during his period of working experience: many of us covered what seems to have been, looking back, an acreage of buff drawing paper at school; we became exponents of elaborate slide-rule mystique; we could consult the Carnegie handbook with one thumb while operating a pencil with the other hand; we would as soon (as one professor who liked to employ the picturesque in his analogies put it) have been caught without our trousers as without a steam-table in our possession.[3] Not so much importance is given today to such ancient virtues. A wide range of new and sophisticated skills, more specialized as to branch of practice, exists. Computer-aided design (CAD) has largely dispensed with the buff paper; factual information is likely to be drawn from the electronic memory. The acquisition or the polishing up of skills is not likely to be closely assessed as a constituent of experience for registration; that it exists will probably be taken as implied by the job description and concurrent membership in a professional engineering association.

The aspect of experience prior to registration we have spoken of as familiarity with industry and commercial customs is similarly likely to be considered as implied in the job description; probably the important thing is that the activities devoted to this benefit be not excessive. There are induction courses in some large industrial companies where service in a succession of departments is the routine, where the process is not confined to engineers in training or those eligible for training status, but includes persons of comparable education in non-allied fields, where the process is directed by a personnel department, and where final assignment to a particular department is by aptitude discovered or by the new employee's choice. Such courses are likely to be carefully scrutinized by a registering authority as warranting acceptance as engineering work, even if the final election be engineering. Where induction courses are under the aegis of the engineering department and are undergone prior to occupying positions in engineering, the situation is different. In industry, the engineer works in collaboration with departments other than his or her own; the engineers in training who accept that the code of ethics applies to themselves before registration just as it will afterwards owe "fairness and loyalty to associates" and must acquire a reasonable knowledge of who they are and what they do.

What has so far been said in this chapter indicates that, in the opinion of the authors, the most important element in an experience record is that it shall show substantial working contact with one or more professional engineers. There may be conventions followed by some regulating bodies that amount to something closer to a definition on this ground than "substantial contact," but we are unaware of it; considering the enormous range of engineer-in-training activities, detailed consideration must be given by the regulating body to each experience submission before a judgment can be made on its acceptance, partial acceptance or rejection.

Partial acceptance simply means that some allowance has been made for what has been submitted, leaving an outstanding period to which some condition may be attached for the guidance of the applicant. A rejection will presumably be accompanied by advice to the applicant as to the likelihood of the employment's ever providing eligible experience. Rejection does not impose a stigma; if the submission reveals contravention of the Professional Engineers Act (that is, unlicensed practice of professional engineering), the applicant will of course be warned to cease and desist; but there is an element of the traditional confessional in the engineer's provision of information, and it seems impossible to believe that a document so submitted would be used as evidence for prosecution connected with an offence committed prior to the submission and the issue of the warning arising from it.

An engineer-in-training, as the term is used here, based on Ontario custom, is a person who, having applied for this status or for membership, is educationally eligible but has not fulfilled the experience requirement. The status cannot be held for more than something of the order of five years (in Ontario). Nevertheless, provided the engineer's education continues to meet the requirement, the status can be renewed with a new application; if the educational requirement has changed at the time of such a new application, the candidate may have to update what was previously acceptable. The point is that rejection of an experience submission does not preclude subsequent steps towards registration. This limited time as engineer-in-training affects persons who may (often for good, sometimes for altruistic, reasons) plan to delay entering engineering work after graduation. It is not necessary to apply for engineer-in-training status immediately after graduation. The regulating authority should be consulted when such plans are being made.

"Substantial" can be a mighty insubstantial adjective; whatever it is applied to will have to be located by someone as a point in a scale of acceptability. Extremes in professional contact in engineering work are easy to imagine: from the safe circumstances of the engineer-in-training employed by a firm of consulting engineers or in the properly constituted engineering department of an industrial concern, we go to the difficult case of one who is doing engineering work entirely without professional tutelage[4] which might occur legitimately, for instance, in Ontario, in the execution of works under Section 12.(3)(a)(c) and (d) of the Professional Engineers Act in a family business. The intermediate range, in which many experience submissions lie, represents occasional, indirect or remote contact with professionals. Applicants who find themselves somewhere in this intermediate range, and forewarned, should set about improving the closeness and frequency of such contacts. They should participate to the full in whatever extramural activities of engineers they are entitled by their training status to attend, whether these are meetings, delivery of papers or bean-feasts, for often, when candidates seek references from professional engineers to accompany their experience submissions, they find themselves all but unknown to those they approach, and can get no more than a sort of hearsay recommendation. More important, the engineers within the candidate's own company should be visited periodically as a tactful reminder of the engineer-in-training's presence and of the fact that the candidate will be citing one or more of them as referees. There is no conscientious professional engineer who will not take this seriously (indeed, the training of successors is an ancient obligation of professionalism, as old, at least, as the Hippocratic Oath) and keep a weather eye on the engineer-in-training's work and progress so that the professional may speak knowledgeably when the time comes. Even where there are no engineers employed, a request should be made of the management, with an explanation of the reason, to participate in dealings with consulting engineers where these are retained. If, as is likely, the engineer-in-training has been taken on the strength of future engineering potential, the employer if he understands the necessity for it, should co-operate in this in his own interests. One way or another, remote contact has to be maximized, and there is reason to believe that many applicants whose registration has been delayed on grounds of experience have not made thoughtful efforts to bring this about.

Decisions on experience gained abroad may be deferred for lack of adequate information; it is difficult to visualize, from a few facts, circumstances where firms, persons and customs are not known. In proportion to the risk of such delay, applicants should be meticulous about what is provided. An original testimonial of an employer showing dates and the title of the position held is desirable; the nature and magnitude of the employer's operations may be quite unknown to the registering authority and may have to be manifested by published matter; information about engineers cited as contacts with the profession will have to be complete enough to enable judgment whether these possess acceptable qualifications as professional engineers, since there are many parts of the world, including some states of the United States insofar as employee engineers are concerned, where a non-engineering person may be doing legally work that would be reserved to professional engineers in any of the Canadian provinces.

Engineers frequently move into areas of activity that are not, under the law obtaining in their jurisdiction, or to be inferred from it, reserved to the professional engineer. Generally, these will have qualified for registration during a period of conventional engineering employment; the change, perhaps gradual, has occurred in relative maturity. There are, however, those who, immediately after graduation, or at any rate before the stipulated experience period has been completed, enter such employment. An experience submission under these circumstances will presumably face two questions: is it engineering work and are professional engineering associations connected with it?

As to the first question, the border between engineering and pure science, applied mathematics, aspects of modern management, financial administration and so on seems to be increasingly indefinite, and strict adherence to the range of work reserved by law to the profession in judging what is engineering work is probably unlikely; in fact, if the second question is adequately answered—that is, if professional engineers are widely participating in the sort of work involved and the applicant can show substantial working contact with these—it seems fair to suppose that a reasonably good case exists. Perhaps pure management occupations are the most doubtful and these should be most carefully and without exaggeration outlined in the submission. Avoided should be the style employed in a case made known to your authors of a young real-estate development manager (of identity not divulged to them) who offered as evidence for registration that he was on "a fast track promotionwise" and "hired engineers when they were needed"!

In this matter of interpretation of occupations not firmly within the scope of engineering, the applicant at least can be confident that the registering body to which he submits his experience resumés is not less modern than himself in awareness and outlook.

NOTES

1. The first recorded move to establish engineering as an educational discipline was in 1775 when the now justly famous Ecole des Ponts et Chausées was established. The transfer of engineering education to the schools was slow. This arose:

   . . . in part from the ill-disguised hostility of those having to do with traditional literary and philosophical studies [but] rather more from the fact that the great majority of practising engineers, wedded as they were to the pupilage system, did not look upon the acquisition of theoretical knowledge as an essential part of a young man's training. The ground taken was that engineering was an art rather than a science. The educational foundation given in the schools of engineering was therefore deemed to be only mildly auxiliary. Even men of the eminence of Rankine and Fleeming Jenkin were sometimes spoken of as "hypothetical engineers."

   C. R. Young, *Early Engineering Education at Toronto, 1851-1919* (University of Toronto Press, 1958).
2. C. R. Young was an eminent Canadian civil engineer, who was on the teaching staff of the University of Toronto from 1908 to 1949, for the final eight years as Dean of the Faculty of Applied Science and Engineering.
3. The story (no doubt apocryphal) is that the professor, further on in his lecture, asked for the loan of a steam-table.
4. The word "tutelage" is taken in its oldest sense (ca. 1650) as "the condition of being under protection or guardianship." (*Oxford*).

CHAPTER 4

THE ENGINEER AS VICAR

HUGH ADCOCK, P.ENG.

Vicar: one who takes the place of or acts instead of another.

The Oxford Universal Dictionary

There is an obvious similarity between the role and duty of an employee engineer as a faithful agent or trustee of the employer and the role and duty of a consulting engineer as an adviser to and, at times, agent of the client in professional engineering matters. Employees' roles are not always as clearly defined as that of the consultant because so much of their work is in-house; in their engineering they are faithful agents, not only of the employer, but of their fellow employees also, whose safety, comfort, productivity and general welfare are affected by the engineers' exercise of ingenuity, their design decisions and their care that the decisions are carried out. In dealing with other than fellow employees their position is similar to that of the consulting engineer. In both cases the engineers are in a position where they are the visible presence of the owner in the direction and control of a project as it is carried through its different stages by the contractor. By substituting "master mechanic" or "production superintendent" for "contractor" and thinking of the mill or factory as the site, we envisage the similarity in the philosophy of those relationships, be it factory or site work.

Factory or office activities of an engineer usually involve long-term day-to-day association with a relatively permanent group which includes members of both management and labour. A field engineer often must have and must exercise special talents to deal with and reconcile the sensitive relationship between the contractor and the engineer. Whether it be in the factory, the office or the field, the talents that succeed are much the same.

One of the more difficult problems the field engineer has to face is his personal relations with the contractor when he is the engineer-in-charge of a project. It is very hard to strike just the right note in personal relations. Some men are born with this "knack of leadership." They never have to argue. They never raise their voices. They say "do this" and it's done. They have the respect of their associates and run a job well. Conversely, some engineers do not have this knack and never learn it. They find it very hard to get contractors to obey instructions. They are either too easy-going or too rigid. The result is constant bickering and lack of harmony. The job and the owner suffer.

Just what should be the attitude of the engineer toward the contractor? How can a balanced, harmonious relationship be attained—and maintained?

There are no hard and fast rules. Individuals vary, as do jobs and contractors. The contractor is in business to make money. The engineer's task is to see that the job gets done and done right. These different viewpoints are not necessary incompatible.

It is well to realize, and it should not be too difficult to convince the contractor, that all job forces are on the same team. The owner wants the best job he is entitled to, at the earliest practicable time. The engineer and the inspectors are there to get this job done for the owner. But the engineer is also an arbiter and must resist any pressure on the contractor to do more than the contract calls for or to do extra work without fair compensation. You will find that most contractors want to build and maintain a reputation for good work.

In the interest of better engineer-contractor relations and to help young engineers who are just starting in the profession to develop leadership, the following six points are worthy of your consideration.

Be Firm

Once you have made up your mind, stick to it. Let's assume that you have thoroughly thought out a certain situation and have made a decision. You tell the contractor that something must be done in a certain manner. He starts raising the roof. Don't let him scare you. Nine times out of 10 he is screaming for effect—or just to see how serious you really are. If you let his noise bother you, or change your mind, you are in for a lot of the same treatment every time. Make sure you are right. If you think you are, stick to it. But if some time you really are wrong, be man enough to admit it and correct your error. You will not lose standing by being fair.

Don't Let Anybody Rush You

Many times you may be asked for a quick decision. Don't be hurried. It's best to take the situation back to the office with you and think it over in all its ramifications. You can be sure that the other fellow has thought it over. Ask yourself if this change or decision you have to make affects only what you are doing now, or will affect something else later. Remember that you may be setting a precedent. Nothing looks quite so bad as changing your mind once you realize the full implications of your snap decision. You can't tell the contractor one thing one day and another thing the next.

Be Fair

Many young engineers, and older ones too, have a tendency to apply specifications too rigidly. They "go by the book." Even when they are shown an equal or better solution to a problem, they insist on following the specifications to the letter. In situations like that, common sense is your best friend. Your judgment will tell you when your contractor is honestly trying to do a good job. Remember, many contractors have been around a good many years and have had a lot of experience. It pays to listen. There will be lots of times when you will be glad you did.

Don't Be Overfriendly

Many contractors try to cultivate the engineer. They are very eager to please him and to become good friends. Starting with coffee breaks, you may go to lunch and have a drink, and before you know it you will wind up by being very, very chummy. You'll be on a first-name basis. Soon you'll start thinking that the contractor is a fairly good Joe. Quite probably he is a very pleasant fellow.

Now you are in something of a predicament. Because when the situation comes up—and it surely must—it's awfully hard to tell one of your "friends" to do something you know very well he isn't going to like doing. You are going to be thrown together often during the course of your project. And no one wants you to be stiff and unfriendly. A pleasant, courteous attitude with a modicum of reserve will help maintain an easy businesslike relationship. However, you should not let friendliness distort your judgment and lead you into "friendly," biased, and unsound decisions. Keep your friends for your personal life.

Think Ahead

It is taken for granted that the engineer is thoroughly familiar with what is happening on his project every day. But how about what's going to happen tomorrow? Try to anticipate tomorrow's trouble today. Look ahead. If you spot any trouble, talk it over with the contractor. Your foresight will probably save money for both the owner and the contractor.

Be a Diplomat

A soft answer gets better results than loud talk. Ask or suggest rather than order. Grease the wheels a little. An engineer acts somewhat as a broker in that he tries to bring both parties, owner and contractor, together in harmony, the end result being a job well done.

For better or for worse you are stuck with your contractor for the duration. It's a delicate situation at best. But with common sense and good engineer-contractor relationships, your owner will get a good job, your contractor will do all right, and you will increase your stature in your chosen profession.

In-house activities of the employee engineer generally involve long-term day-to-day relations with a relatively permanent group, and he shares to some degree the interests of long-term fellow employees. Usually he cannot subscribe wholly to any group ethic—declared or perceived—among them. In the welter of labour laws that prevail across Canada and the U.S.—probably throughout the industrialized world—the engineer is not commonly included by compulsion in industrial unions but is identified with management. There are many instances, no doubt, where a union affiliation is required of engineers, particularly during early stages of learning the processes of an industry that involve periods of work at different jobs prior to engaging in the practice of professional engineering in these precincts. There is no prohibition of such affiliations in the Code of Ethics and it will probably be only in actively adversarial circumstances that young engineers will have to judge their conduct with nicety. To have to face such judgments is the price of maturity for every man and woman.

Professional unions exist in some large organizations where there are many employee professional engineers. These, presumably, are able to conduct negotiations concerning their interests with the employer without contravention of the Code of Ethics. Fairness and loyalty travel two-way streets, and a common interpretation of these virtues can be reached without grievous belligerence.

For employee engineers in public service and, with variations, in the armed forces, a patchwork pattern (federal, provincial, municipal) of what amount to contractual obligations, sometimes accepted under oath, may impinge marginally upon the Code of Ethics (as in the matter of secrecy versus Section 8 of the Code) but at the sophisticated level of practice at which such conflicts might appear, reconciliation of both obligations seems possible.

For further information on those special aspects of ethics that are related to professional engineering practice in public service, civil or military, your authors recommend the book *Ethical Conduct: Guidelines for Government Employees* by Kenneth Kernaghan. Published in 1975 by the Institute of Public Administration of Canada, the book presents an excellent treatment of the subject. In its foreword it is stated that "the Institute is specifically enjoined by its Constitution to concern itself with questions of public duty and professional etiquette."

The guideline deals with these questions in Canada in the federal, provincial and municipal governments and makes comparison with the nature and treatment of the problem in the central governments of Great Britain and the United States.

CHAPTER 5

PROFESSIONAL ENGINEERS IN THE MANUFACTURING INDUSTRY

HOWARD BEXON, P.ENG.

While the engine runs, the people must work; men, women and children are yoked together with iron and steam. The animal machine is chained to the iron machine that knows no sickness and no weariness.

Hammond and Hammond, *The Rise of Modern Industry*

This bitter passage is quoted by the Hammonds from writings of the late 1700s, that darkest period of the Industrial Revolution that followed the adaptation of the engine to the driving of rotating shafts, the "fateful combination of fire and the wheel" by which our technological society was born, and of which the misery and sorrow were only gradually abated in the ensuing half-century.

Watt died in 1819 in his 83rd year. There is a portrait statue of him in Westminster Abbey, inscribed:

JAMES WATT
who, directing the forces of an original genius,
early exercised in philosophic research,
to the improvement of
THE STEAM ENGINE,
enlarged the resources of his country,
increased the power of Man,
and rose to an eminent place among the most
illustrious followers of science and the real
benefactors of the world.

For generations these two conflicting philosophies have exercised moralists in our society. At what level of immediate threat to the welfare of the human race shall engineers and scientists stay their hands? And if they do so, to what extent may they be failing to become "real benefactors of the world"? Yet, in practical life, the questions have a certain unreality. Given human nature, neither discovery nor development will be locked away in the hope that some era of perfected humanity will arrive. In one of the scenes of a play concerned with such speculations, God is told that people from Planet Earth had landed on the Moon. "Ah, yes," He said, "I was there. It was most gratifying. Man has been so slow to learn about the world I put him in charge of. After all this time he's beginning to realize what he can do with the tools I have given him."

The foot of humankind is set on an irreversible path, and if we have traded off something of the immemorial concepts that gave dignity and stability to the individual, accepting in return a philosophy of endless economic growth and endless increase in the material standard of living, we can only strive to organize our world so that we may exist in reasonable harmony with our destiny. Having all but forgotten fear of God, we must learn to meet the new fears that beset us.

Canadian Technology in Manufacturing

Canadians live under the psychological dominance of the United States. We want their "standard of living," indeed, feel we deserve it in spite of obviously different circumstances. To do so, we have to import most products and much production technology; we simply cannot afford to do it ourselves because our market is too small. The cancellation of the Avro Arrow in the late 1950s brought this crushing reality into focus. Less obvious is the foreign or government ownership of much of our large manufacturing industry such as aircraft, automobile and appliance. We do not lack the ability (the Arrow was potentially great) or the desire; we just cannot afford it.

Despite this, Canada poses some unique problems of climate, geography and culture that other countries do not have to the same degree. In these areas, our technology has flourished and we are able to export this to fill the weaker needs of foreign countries. In "bush" aircraft, snow removal equipment, railway products, snowmobiles, bridges, large-scale farm equipment, Canada is a leader in both design and manufacture.

New generations of youthful engineers and keen immigrants from foreign lands are adding a new impetus that will bring fresh vigour to Canada. The important thing is that we exercise wisdom and restraint to attack appropriate problems with appropriate engineering.

The Role of the Professional Engineer

We are fortunate that engineering is a recognized profession in Canada, unlike many other countries, and professional engineers are in a powerful and responsible position in our technological society. Eighty-five percent of them are a part of large organizations, most generally manufacturing establishments. Their established ethics, as members of a profession, is valuable security to any employer. The same code that establishes the long-range goals of the professional engineer is either complementary to or the same as the employer's. As a result, engineers are in the most diversified and far-flung of all the professions and at least half work in jobs with no relation to design at all. The only other profession approaching this situation is accounting.

Let us now examine the role of professional engineers in this picture. Obviously they have a major part to play in nearly every aspect. Less obvious is the fact that they have a greatly different relationship to their employer than do their colleagues in private practice. Private practitioners enter into a three-way enterprise to provide service, together with the owner of the work and contractor who will build it. The owner and professional engineer have one contract, the owner and contractor a separate contract. Thus the engineer in private practice serves the interest of the owner. On the other hand, the employee engineer is in the service of the "contractor" or manufacturer and seldom has any relation at all to the "owner" or purchaser of the product. By the principle of vicarious liability, the employer is liable for the negligent acts of employees; thus the responsibility of professional engineers as employees is to protect the employer, rather than the owner, from harm. The end result is the same, yet the employee engineer may have an even greater responsibility in the sense that manufactured goods are produced in such large quantity and manufacturers have not only substantial wealth, but a great concern for their corporate responsibility, and are particularly vulnerable to immense liability lawsuits. This is applicable also to the process of manufacturing where similar crushing liability suits arise over safety, environmental and social issues. In all this the need for powerful, proven ethical standards is evident; the already proven ethics of the engineer is a near perfect match to the need.

Product

This is the classic work of engineers. Although the concept may or may not be theirs, they are responsible for the efficient, adequate design and for definition of the loads, performance and safety requirements. They carry out the scientific task of calculating performance and strength, proving the product prototype in shop tests and later field tests and repeating each step by iteration until optimization is achieved. This process involves a continuing participation in improving input design parameters (with users, specifying bodies, and so on) and in matching the design to the most efficient available production processes. It also involves working with other manufacturers and agencies, such as standards organizations, since a new development is seldom carried by one firm.

The product function is very broad and encompasses the entire life cycle from research and development, through initial production, continued improvement over its life and responsibility for quality control, reliability and field servicing. The mature firm handles this through a series of departments such as research and development, product engineering, test department, quality assurance and customer service. Typically, professional engineers are employed to carry out and/or supervise and manage the work of all aspects.

Manufacturing

The design and construction of the industrial process is the task of the industrial engineer, as is the continued improvement that is so imperative in a competitive nation. Similarly, selecting the manufacturing process and designing the tools for producing new products and components is a part of this task. An offshoot of this, of crucial importance, is the procurement of materials and complex planning of schedules to assure an efficient and orderly flow of materials into and out of the factory. A second offshoot is the responsibility for safety within the plant and protection of the environment from pollutants created in the process. The maintenance of the equipment in efficient working order affects almost all aspects and is a further major responsibility.

Traditionally, the leadership of manufacturing is generated from its own ranks, and this is still frequently the case even in large organizations. However, the increasing complexity of the production process and the enormous cost of larger installations has seen a trend toward the engagement of professional engineers, particularly at levels of supervisory responsibility. A large part of this is due to increasing technical complexity of the equipment. The broad educational base of engineers gives them not only a basic understanding; they are also equipped with a methodology (the "scientific method") to tackle new and abstract problems that the shop worker does not have. Since future corporate management will be formed from these ranks in many cases, the trend toward hiring engineers is continuing, and accounts for a large part of the 50% of engineers whose work has little conventional engineering content.

Marketing

Without sales of its products, the company cannot exist. Usually professional engineers have nothing at all to do with sales of consumer goods, but they have much to do with the sale of technical products, particularly where the "customer" is another engineer or technologist. In general the more complex the product and its application, the greater the involvement of the engineer in its promotion. Marketing includes direct sales, advertising, application studies, negotiation of contracts and preparation of proposals (plus the occasional lunch). It may also include establishment, training and servicing of distribution networks. "Sales engineering" is another promotional tool, and a good one, in which engineering service is provided by a product manufacturer to assist customer product engineers in using the product effectively and to inform them of the capabilities of a product.

No two organizations are constituted in the same manner. This is not only because of their different products and position in the life-cycle of their product; it also represents their "corporate strategy," the philosophy of management and the owners. Profound influences also are imposed by geography, social values, government policy and the labour force available. Therefore, one does not find the divisions clearly established, nor is the employment definition of the professional engineer predictable.

Further confusion is frequently added by the tendency of manufacturers to apply the title "engineer" to employees who lack professional status. These people, often shop-trained and often competent in a narrow field of endeavour, are employed in engineering positions by their company, without professional status, in all facets up to and including product design. This situation is really a liability employers take on themselves. It is only of concern to the profession if such individuals carry the title engineer into the public arenas since the public is unable to discriminate between "engineer" and "professional engineer."

The Importance of Ethics in Manufacturing

Obviously, professional engineers in manufacturing are in a dynamic, evolving milieu as the corporate form continues to be reshaped under so many influences. The notable aspects are that more and more positions are being offered, the status of the employee engineers is rising markedly and their responsibilities are spreading horizontally and vertically throughout. This in turn is affecting both the concept of the profession itself and the nature of the engineer's education, which is being pressured toward more management training as well as technology.

It is likely that most employers are either unaware or only vaguely aware of the professional engineer's code of ethics. This indicates that the engineers themselves, through their image and behaviour, are earning the respect that is elevating their status and demand. The prime importance of the code of ethics therefore is the effect it has in shaping the attitude of engineers towards themselves and their colleagues.

An understanding of history will provide engineers with respect for their forebears and a strong respect for all other trades and professions. There is no reward in industry for arrogance. This same historical perspective will explain many questions regarding the status of the employee engineer in Canada. Above all, it sets a challenge for our future, and the future of Canada.

When professional engineers, through training, experience, and by projecting the dignity inherent in their profession and ethics, have demonstrated their potential to contribute to the manufacturing community, they and their colleagues in that group will receive the same level of respect as is afforded the private practitioner. Public response will follow; indeed this is now clearly happening.

CHAPTER 6

RESOURCE INDUSTRY— THE ENGINEER AND THE APPLIED SCIENTIST

GEORGE ADAM, P.ENG.

Prospectors and Explorers

In resource industry, as we contemplate it here, we see at least two broad phases: first, discovery and assessment of value, quantity and accessibility of what is to be exploited, and, second, the ongoing operation of processing the yield until a marketable product is obtained. The resource may be non-renewable or, as with a stand of prime timber, renewable only over many years, and in both cases many of the concerns we are about to express apply. But, although the latter, the slowly renewable, are susceptible to management methods that should take their development out of the class we call extraction operations, we do not limit ourselves here to the latter, i.e., discovery and processing of ". . . oil, natural gas, coal, metallic or non-metallic minerals, precious stones, other natural resources, or water . . ."[1]; in such case we imagine the forester and the agronomist as the applied science professionals, in lieu of geologist and geophysicist.

Evaluating the Find

In the first phase, discovery and assessment, the geologist and the geophysicist are themselves, or they follow, the prospectors; they are likely to be the first professionals to form judgments and make recommendations to a client or employer; they are armed with sophisticated instruments and knowledge, yet they have something in common with Columbus seeking a Western route to Asia: the great Genoese was answerable to Isabella of Spain, as our professionals are answerable to employer or client. But the world of the 15th century imposed no obligations upon its explorers and developers to have a care for the public interest and welfare. Nature still seemed of such vastness that it would be affected by man's acts about as much ". . . as the sea's self would heed a pebble cast."[2] Today, the geologist, the geophysicist, and the engineer, presumed to have become a participant in the planning of the step by step development of the resource, must bear the solemn duty of protection of the public from harm in any decisions taken. In the extreme, this duty may comprehend a consciousness of peril to the very existence of mankind.

Future Adverse Effects

In the more immediate aspects of the public interest—assurance that the cost of protection to, or restoration of, the local environment is provided for in the feasibility estimate (considering the environment as the social as well as the physical ambience)—as well as in the long view comprehending ulterior effects, the professional, be he engineer or scientist, must treat his concern for the public welfare as one with his loyal service to his employer; in protecting the one he is protecting the other against falling inadvertently into the role of plunderer. For the activities of a resource industry are more often than not carried out far from the public eye, and the professional who is privy to the planning of such activities becomes, as it were, an advocate of the public. This is the prime obligation of any man or woman who follows an occupation reserved by law to persons of certain qualifications. Such reservation, long ago intended to protect the public against bodily accident, fraud and incompetence, gave rise to the professions.

Confidentiality Is Essential

The professional is not, repeat not, a private eye whose duty it is to place information about the planning of an enterprise before the public, making the media the arbiters of action. He is the advocate of the public interest in the councils of the enterprise that is his employer or client, and one of his obligations is confidentiality. He must determine in his own ethical conscience and judgment what is acceptable as environmental effects; he must guide and convince his employer to plan within these constraints of acceptability.

Alberta and the Northwest Territories

Reverting to the extraction of petroleum and mineral wealth, where the engineer, the geologist and the geophysicist normally collaborate, there are, at the present time, two jurisdictions in Canada, Alberta and the Northwest Territories, in which these three professions are regulated under one law, in Alberta, The Engineering, Geological and Geophysical Professions Act and in the Northwest Territories, an Ordinance. The merits of this form of regulation are being considered by other jurisdictions as they review and update their legislation.

In Alberta, for some 70 years, the position of the geologist and geophysicist in relation to that of the engineer was dealt with legislatively, but in a varying manner: specific reference to the application of geology in the definition of the practice of professional engineering appears in the Act current in 1922, but it was an amended wording in 1942 which included ". . . geological and other scientific investigations . . ." of rocks, mineral deposits, etc., that suggested relevance to geophysics.

Uneasy Unity to Independent Recognition

The 1943 revisions to the Engineering Profession Act, 1930, made it possible for geologists and geophysicists with satisfactory qualifications to be registered and designated as professional engineers. Although this situation obtained for some 17 years, geologists and geophysicists still sought professional registration as such, rather than as professional engineers. In 1960 a major step was taken: an Act "to regulate the professions of engineering, geology and geophysics" came into force, entitled The Engineering and Related Professions Act. This Act introduced definitions of the practice of each of the three professions, as well as introducing P.Geol. and P.Geoph. as titles independent of the P.Eng. The current Act is The Engineering, Geological and Geophysical Professions Act, 1981, with amendments to 1984.

Acknowledgment

We are indebted to APEGGA's publication *The Practice of the Professions of Geology and Geophysics* for information from which we have extracted these few highlights with which we have done less than justice to what must have been an arduous labour of developing legislation spread over years. The preservation of the three professions as entities united by a common discipline and subscribing to the one common Code of Ethics must be a reassurance to the resource industries against dangerous or unscrupulous practices.

NOTES

1. From the current Alberta Act.
2. Fitzgerald's *Omar Khayyám*.

CHAPTER 7

STANDARDS AND CERTIFICATION

JOHN KEAN, P.ENG

A unit is a particular amount of the physical quantity to be measured, defined in terms of a standard.

U.S. Bureau of Standards, Circular No. 60, 1920

Standards for physical quantities do not normally give the scientist and the engineer much concern; to them, litre and gallon are volumes expressed in the dimension L, and if the units of L are in terms of the length of a metal bar cherished under glass, or of some fraction of the circumference of a great circle of a perfect terrestrial sphere, or of the wavelength of the electromagnetic radiation arising from the optical transition (in the orange range) in the isotope of krypton of mass 86, so be it.

Engineering design would be primitive without recognized standards for physical quantities. Our forefathers invented standards from time to time, or adapted existing standards. For example, the working capacity of a London draft horse was eventually expressed in terms of existing standards of force, length and time by Watt to give us a standard for power from which we derive nowadays kw, k cal/h, and so on by simple arithmetic, the draft horse having been relegated to history, except for his place in the word "horsepower." Today, it is only when we steer into distant and untravelled seas of scientific knowledge that a new basic standard is needed.

Arbitrarily calling a basic standard one that can be divided into units of quantity, one might propose a thesis that standards of quality or suitability in materials, functions of systems, workmanship, perhaps even beauty, could, in some ultimate analysis, be measured by evaluating a series of constituent standards of quantity by test. Such a thesis probably could not be sustained: go/no-go tests (ultimately, it works/does not work) do not yield measures. In our context the speculation is idle, for it does not illustrate the important connection between established standards and the ethics of design. In this connection, Section 86.-(2)(h) of O.Reg. 538/84 states, "For the purpose of the Act and this Regulation, 'professional misconduct' means . . . undertaking work the practitioner is not competent to perform by virtue of his training and experience."

Seldom would a design project be entirely within the competence of one engineer or a group of engineers if there were not available a multitude of existing designs for components and subsystems of which the performance and suitability is guaranteed by others; the designer's competence under the code of ethics now lies in the training and experience by which the designer judges the knowledge and authority of those who give such a guarantee. Machine designers, for all their expertise, may not be really at home with electrical connections, but they do not have to retain an electrical engineer to specify the plug that will mate with domestic wall sockets. Their conscience and their legal obligations are satisfied because they accept the recommendation of a certifying body of recognized ability and integrity and so specify the component. They are accepting a standard of quality and suitability; in this example, a fitting about the size of a walnut, familiar to every homeowner, has been given minute study already; our machine designers need lose no sleep over it.

The International Organization for Standardization (ISO) defines a standard as:

> A technical specification or other document available to the public, drawn up with the co-operation and consensus or general approval of all interests affected by it, based on the consolidated results of science, technology and experience, aimed at the promotion of optimum community benefits and approved by a body recognized on the national, regional or international level.

The complexity of standards of quality is not always appreciated by the public for whose protection the standards have been defined; the indignant shopper who protests the difficulty arising from the advent of *mL* instead of *fl. oz.* is likely to find the labelling of a can of peas as *Canada Grade A* perfectly straightforward—a good example of our human propensity for straining at the gnat and swallowing the camel.

Certainly, standards that deal with materials, products, structures and systems are complex and time-sensitive; they are subject to frequent change to keep up with our fast-moving technology, environmental concerns and the future availability of materials. This is the type of standard that is the preoccupation of standards-writing organizations around the world.

However, in its simplest form a standard is no more and no less than a way to deal with a problem, a problem then can recur and if left unresolved can cause waste, unsafe conditions and dissatisfaction. The degree to which the standard solves the problem depends on the ability of those working on it to satisfy the various parties who may be affected. The formalizing of the process in a set of rules leads to the publication of a formal standard.

In Canada the need for formal standards arose during the First World War when a group of Canadian engineers and industrialists realized that if Canada were to progress as an industrial nation it needed formal engineering standards. The lack of such standards had, in many instances, seriously hindered the war effort.

Application was made to the federal government for a charter to establish an organization to develop national standards for Canada. The result was the creation in 1919 of a voluntary non-profit organization known as the Canadian Engineering Standards Association. Since then this organization has extended its activities well beyond the engineering field; in 1944 the name was changed to what it is today—the more appropriate Canadian Standards Association, or CSA. Other standards-writing organizations have also come into existence in Canada to serve specific needs.

Here is a partial list of American, Canadian and international standards organizations whose activities affect Canadian practising engineers in many ways:

American National Standards Institute (ANSI)
American Society of Mechanical Engineers (ASME)
American Society for Testing and Materials (ASTM)
Bureau de Normalisation du Québec (BNQ)
Canadian Gas Association (CGA)
Canadian General Standards Board (CGSB)
Canadian Standards Association (CSA)
Institute of Electrical and Electronics Engineers (IEEE)

International Electrotechnical Commission (IEC)
International Organization for Standardization (ISO)
National Sanitation Foundation (NSF)
Society of Automotive Engineers (SAE)
Underwriters' Laboratories of Canada (ULC)

We can scarcely imagine engineering design and procurement without their standards, either in domestic practice or in international commerce.

Standards-Writing in Canada

Because it is such a costly operation to develop a standard, careful consideration is given to ensure that there is truly a national need. However, once a decision to go ahead is made, the next step is to ensure that a properly constituted committee is established, one that is representative of all interests that may be affected by the standard; designers, manufacturers, purchasers, regulatory authorities, users and technical experts all become part of the process. In establishing the matrix of each committee, we try to provide a balance of interests so that no one individual group can unduly influence the process.

Another feature of Canadian standards development is that normally it is done on a two-tier basis—the technical committee responsible reports to an overview committee, which has the responsibility to ensure that the final document will not conflict with already published standards for similar or allied equipment and that proper steps have been taken and proper procedures observed, particularly with respect to the handling of negative comments.

In the development of a national standard under the Canadian system the consensus principle is used. That is to say, decisions are made through the reconciliation of diverse opinions until substantial, but not necessarily unanimous, agreement is reached within the standards committee.

To ensure that maximum use is made of what already exists, when starting a new project most committees will first look towards the international scene to see if a document is already available that may be adopted in its entirety or in part as the basis for the Canadian standard. Usually when an international standard is not available or is not suitable, we tend to look south of the border to our American friends because of the existing linkage between our two economies. The process is not always a one-way street: our counterparts in the U.S. often, in starting their standards projects, look to see what has been developed in Canada. In fact, there is effective liaison among standards organizations throughout the world and especially between those in the U.S. and Canada.

With the concept of global markets, Canada is becoming increasingly involved in international standards work. To ensure that Canada is effectively represented in international standards forums as well as to co-ordinate standards activities in Canada, the federal government in 1970 created the Standards Council of Canada (SCC). Under the SCC, Canadian standards-writing organizations can become accredited and take part in the formal process of having their standards adopted officially as National Standards under the procedures of the National Standards System. This provides a further check on the development of consensus standards to ensure that due process has taken place with the added feature that all National Standards are available in both English and French.

Because we are working under the principle of consensus, not everyone will be satisfied with the end product, and often to reach a satisfactory conclusion, trade-offs are necessary. This can sometimes present concerns to engineers, particularly when we consider that under the Code of Ethics professional engineers must regard their duty to public welfare as paramount. In some cases, the most stringent standard may not be the best one, particularly when we take into account a cost-benefit analysis or how the product might be used. For example, when the standard for hockey helmets was developed a few years ago, it was recognized that in order to be useful the standard had to allow for the production of a helmet that would be attractive and comfortable to the user as well as being relatively inexpensive so that it would be used by the majority of people involved in hockey. Yet at the same time it had to protect the user and provide a reasonable degree of safety. The question of what is reasonable as opposed to what might be the ultimate can produce many hours of debate. There can be no easy set of guidelines to determine what is reasonable and each case must be examined on its own merits. In the case of the hockey helmet, the standard did its job and the product that is now being produced to the standard has all but eliminated certain injuries previously reported.

Engineers may also be put into a difficult position if they are in serious disagreement with other engineers on a committee. This problem can be compounded if, at a later date, the disagreeing engineers are asked to design or build in accordance with that standard after it has been published. A similar situation arises when engineers realize that their votes on a committee in favour of a particular standard may make the designs or products produced by the employer obsolete. Even if they are not made obsolete, added costs and changes may be such that the employer is placed in a situation where there is temporary or even long-term loss of market. Again, there are no easy answers.

Some comfort may be taken in the knowledge that standards-writing technical committees are made up, in large part, of professional engineers who are undoubtedly wrestling with the same matter of reconciling their decisions with the Code of Ethics. In any event, the committee member has to be satisfied that there has been no negligence in reaching a conclusion and that all points of view have been considered. It should be remembered that where an issue is controversial, there is often no right or wrong position; it is merely that there is more than one opinion as to the optimum.

In Canada today, the largest single group active in the development of standards is the engineers; probably this is true also for most of our trading partners, depending on what is accepted as the legal definition of an engineer in the country being considered. Within all of the Canadian standards-writing organizations (SWOs) there are close to 10 000 individuals involved in the standards development process and about 3000 of them are professional engineers. When we consider the direct costs borne by the various SWOs as well as the indirect costs paid by individual companies, government agencies and associations who pay the salaries and travel costs of those who attend standards meetings, a fair estimate of annual cost in Canada is more than a hundred million dollars. This business of standards is indeed big business with a pronounced engineering accent.

Certification

In recent years, the idea of "certification" has gained prominence in the standards world. The concept, however, is not new; in fact, it had its origin in the very foundation of standardization. The first time a craftsman claimed that a product met a commonly accepted standard, the most basic form of certification came into being. Hallmarking, the listing or approval of products, the granting of the use of registered marks or certificates to indicate conformity to a standard, are all included in the general concept of certification.

Changes in social and economic conditions have caused continuing evolution and growth of the certification process.

The costly procedure of developing a standard is futile unless that standard is used. It may be used in any of several ways. In many cases, designers and manufacturers voluntarily agree to follow appropriate standards and their customers accept their word. Here we are looking at a two-party system; the bulk of world trade is carried out on this basis. The customer or second party can, if it wishes, carry out its own audit of what is being produced but still only two individual groups are involved: the producer and the user. Over the years, however, business and trading patterns have become very complex. Quite often the end user has little knowledge of the producer and may wish to involve an independent or third party in the trading process, to act on behalf of the user to ensure that there is compliance with whatever standard is being used. This form of compliance with standards has now become universally known as third-party certification.

In many countries, and this is particularly true in North America, not only do purchasers require independent audits of manufacturers, but governments do also, especially when public health and safety are involved. For example, nearly all provinces in Canada now require that electrical products sold and/or installed within their respective boundaries must bear the CSA certification mark for their sale to be legal and for their installation to be approved. Similar regulations apply for fuel-burning appliances and other products.

In Canada, three of the major standards-writing organizations—CSA, CGA and ULC—offer certification services to their standards. Also, these three organizations provide certification services to other standards within their respective scope of activities. CGSB announced recently that it also intends to offer a certification service to its standards. Because of the close relationship between standards, certification and testing, the Standards Council of Canada also offers accreditation services for organizations engaged in these latter two activities. In addition to accrediting CSA, CGA and ULC as certification organizations, the SCC has given formal recognition to Warnock-Hersey Professional Services Limited and the Council of Forest Industries. Details of the National Standards System and the SCC's accreditation programs can be obtained through the following toll-free number from any point in Canada: 1 (800) 267-8220.

Although certification programs do vary from product to product, there are certain similar features regardless of which organization provides the service. Samples are usually required for test and evaluation, and a procedure that has come into use recently requires manufacturers to demonstrate how they intend to ensure continuing compliance on the production line. Pre-licence inspections are becoming quite common, as are complete checks of the manufacturer's quality control and, in some cases, quality-assurance systems, depending on the product in question. Once the certification organization is satisfied that the manufacturer has the ability to comply on a continuing basis, certification is granted. This normally consists of the issuing of a licence allowing the manufacturer to place a certification mark on all products that the manufacturer claims meet the particular standard on which the certification was based. The manufacturer, as part of the licensing agreement, grants the certifier the right to audit the manufacturer's production as long as the agreement is in effect. Violations of the agreement can result in cancelling of the licence, requests for recall and public disclosure; in some cases, legal action is taken by the certifier against the manufacturer. Where a manufacturer is required to have the product certified prior to sale or installation, the releasing of a sub-standard product violates not only the licence agreement but a legal statute as well. According to O.Reg. 538/84, Section 86.-(2)(d), *inter alia*, "professional misconduct" means "failure to make reasonable provision for complying with applicable statutes, regulations, standards, codes, by-laws and rules in connection with work being undertaken by or under the responsibility of the practitioner."

The success of such a certification program depends on the degree of acceptance of a particular certification mark by the purchaser or by the regulatory authorities. The marks of most of the agencies we have mentioned have been tried and accepted over a period of many years; in some cases the mark is recognized and accepted not only throughout Canada but internationally.

As with any system, there are benefits and drawbacks. The fact that standards take so long to develop can be restrictive; it can retard innovation, particularly in design standards rather than performance standards. However, for every drawback there is usually a positive benefit. A case in point was the development of standards for plastic pipe for use in the plumbing field. Prior to the 1960s, plumbing inspectors across the country would not accept anything but the traditional materials, such as cast iron and copper. The industry in Canada, aware that the potential for its plastic product would be limited, turned towards the standards process to solve the problem. The standards that were developed included requirements and tests to determine the equivalency of plastic to the more traditional materials, and once products meeting the standards were available, the plumbing inspectors felt that they now had something on which to make a judgment. Thus the standard helped to advance the technology. On the other hand, again to do with plastics, the safety standard for electrical refrigerators for many years contained a requirement that the enclosure be made of metal, mainly because when the standard was originally issued no one thought that anybody would use anything but metal for the enclosure. Because the requirement was in the standard, the use of plastics was precluded and it took several years for the standard to be changed and that design constraint to be removed. Under such circumstances, and where product certification is needed for legal acceptance, delay in modifying the certification standard is detrimental to both the industry and the consuming public. Fortunately, ways have been devised to expedite standard requirements for use on an interim basis pending a complete standard revision.

Looking toward the future, with our growing concerns for energy conservation and protection of the environment, the potential for more standards is virtually unlimited. With our technology becoming more complex year by year, there is every likelihood that more engineers will be required to serve on standards committees representing users, producers, regulatory authorities or even consumers. Engineers will certainly have to work with the results of the standards system. Questions of efficiency, reliability, life expectancy, reasonable safety vs. ultimate safety, and ultimate performance will challenge these engineers and present them with unique opportunities not only to make a worthwhile contribution to the Canadian community as a whole, but also to advance the concept of professionalism and ethical behaviour.

CHAPTER 8

INTERNATIONAL ENGINEERING WORK

JOHN STEVENS, P.ENG.

The Beginning

When we consider engineering as "the introduction of change into the conditions of living" we realize that it has been going on since prehistoric times. Also, that there have been international, or foreign, projects since the dawn of civilization.

About 8-10 000 years ago an agricultural revolution started in the fertile regions of Iraq and Syria and an acre of land under cultivation became able to support twenty to two hundred times as many people as before. This freed some of the people from the subsistence tasks of hunting and food-gathering and so they were able to engage in other occupations. Permanent villages grew to cities. Cities require specialists. To the tribal wizard or priest and the tribal chief were added merchants, physicians, poets, smiths; instead of building their own houses or carts or wells, people began to acquire these from specialists.

As wealth and experience developed, people undertook projects too large for a single craftsman. Organization of hundreds and thousands of workers came about. Out of this came a new class of people: the technicians and engineers who could negotiate with a king or priesthood to build a public work, a fortress, an aqueduct, a tomb; to plan the details and direct the workers.

These engineers had to be inventors as well as contractors, designers and bosses: they had to be able to imagine something new and transform the mental picture into a physical reality.

Perhaps all that does not seem to have much to do with explaining how engineers became involved historically in foreign projects. However, consider what was happening in the agricultural revolution.

Agriculture required irrigation and irrigation required central administration. Historians may argue whether empire came first and made possible large-scale irrigation or whether large-scale irrigation came first and encouraged the growth of empire. It does not matter to this discussion. What is important is that expansions of empires took place in the valleys of the Nile, Tigris, Euphrates, Indus and Hwang-ho, and technicians and engineers became involved in foreign projects. Expansion of empires meant that engineers and technicians went into "foreign service" to support both civil and military requirements.

Usually history is taught to us in terms of warrior-heroes like Achilles and Hector and the siege of Troy—but about the same time as Troy was being besieged a nameless genius invented the safety pin. Most of us know about Julius Caesar, but who has heard of his contemporary Sergius Orato, the Roman building contractor who invented central indirect house heating? Thus the engineer in history is more often than not unnamed if not forgotten.

As civilizations grew, incentives associated with the use of engineers in foreign projects also developed. These can be categorized as military, religious, commercial, political and altruistic.

The following examples illustrate some of these aspects.

About 1000 B.C., Assyrian military engineers developed the belfry, or movable siege tower. This grew from the simple scaling ladder, which did not provide any protection to attacking soldiers as they tried to climb a fortress wall.

The fully developed Assyrian siege tower had six wheels, a lower tortoise covered with wood and hides, and a front tower with enclosed stairways and water containers to fight fires. In the tortoise a ram was suspended on chains. This siege weapon was the dominant attack device for about 2000 years, until the cannon made lofty walls useless for defence.

When Sargon II of Assyria invaded Armenia in 714 B.C., he saw a form of irrigation supply not then practised in Mesopotamia. It used an underground tunnel sloping from a mountain source to bring water to the dry plains. It had the advantage of reducing losses due to evaporation. As described by James Michener in *Caravans*, ancient examples of this form of civil engineering can still be seen in Afghanistan. Today agricultural engineering is an important part of Canada's international engineering activity.

The Phoenicians living along the Lebanese coast of the Mediterranean were active engineers in shipbuilding. They developed two styles, one a long, narrow war galley, one a broad stubby merchantman, both using oars.

Engineers today have to be familiar with standards. In Greece, roads began as simple tracks with ruts worn by cartwheels. As time went on the ruts were often dug out and lined with cut stone; in Athens, each track was about 20 cm wide, 7-15 cm deep and the gauge was 137 cm. Roads were provided with sidings like railways.

In Syracuse the gauge was 150 cm. Thus a person who brought a chariot from Syracuse to Athens would find the wheels were too far apart to fit the tracks.

In 331 B.C. Alexander took Deinokrates, an architect from Rhodes, to Egypt and commissioned him to lay out the city of Alexandria. The Romans, during the approximately 800 years of the western empire, 400 B.C. to 400 A.D., carried engineering to the far reaches of the conquered countries as well as bringing engineers to Rome. The Romans built roads for civil communications and to expedite military movements, aqueducts for water supply, merchant ships to bring cargos from Egypt to Rome. As well, in unconquered lands like Persia, Roman engineers were in demand for civil engineering works.

Leonardo da Vinci, perhaps best known for his artistic triumphs such as the Mona Lisa, was military engineer to the Duke of Milan for 18 years.

In the middle ages in Europe, mining and metallurgy developed with Germany taking a leading role. By 1560 A.D. Germany was sending mining engineers to England to mine copper near Keswick, zinc in Somerset, and to set up a brass foundry (1568) near Tintern Abbey. These imported engineers came in some cases at the invitation of English companies, sometimes to set up German companies. During the reign of Elizabeth I, German engineers came to Britain to work on hydraulic problems, including draining mines and providing water supplies to towns.

In the period 1750-1780 English engineers were recognized as world experts and went to France to work in foundries, machine shops and water works.

Why Do International Projects Get Started?

A simple answer—international projects are started in the same manner as domestic projects. Someone needs a service or facility; another person or organization can provide the desired service or facility.

However, on the international scene there are differences in degree or in type from those factors that exist for a purely domestic project.

The reasons that lead to an international project being carried out today are often the same reasons that history has shown to have existed in the past—they include military, religious, commercial, political and altruistic purposes. History tells us that most foreign projects are initiated for rational reasons; unfortunately some are started for irrational objectives or misunderstood needs.

However, lest the foregoing leave you with too pessimistic a view of human nature, let us note that not all foreign and international projects are started for selfish reasons. There have been, and are, many instances of foreign projects being initiated for purposes not simply selfish—not purely military, not just for commercial gain. In saying this we can note examples like those supported by the Marshall Plan of the United States, after the Second World War, which was instrumental in the rebuilding of Europe, and Canada's Colombo Plan, which has provided financing for many countries for equipment and training.

For international projects today, the funding may come from any of a number of sources. Consulting engineers interested in foreign work should be familiar with the operations of national and international financing agencies. Some of the principal ones are the International Bank of Reconstruction and Development (IBRD) or World Bank; International Finance Corporation (IFC), also an affiliate of IBRD; the Inter-American Development Bank (IDB); the Export-Import Bank of the United States (Eximbank); the U.S. Agency for International Development (AID); Canadian International Development Agency; (CIDA) the Asian Development Bank; and the African Development Bank. Except for the last three agencies, all have headquarters in Washington, D.C.

In Canada, the primary source for financing foreign projects is the Export Development Corporation (EDC), an agency of the government of Canada.

In judging whether to grant financing for a project, a lending agency generally uses the following criteria, and perhaps others also.

(i) will the project contribute to economic development;
(ii) is it consistent with development plans;
(iii) is it economically and technically sound;
(iv) are adequate self-help measures being taken;

(v) is there ability to repay; and,
(vi) has adequate consideration been given to financial assistance from other international or private sources?

The two main categories of projects are:

Productive projects that will contribute immediately to economic development, such as power, transportation, ports, irrigation and agriculture.
Projects of a social or development nature, for example, water and sewage utilities, schools, hospitals, housing.

All the lending agencies, national and international, have fairly close working relations; to a large extent all follow the above criteria and priorities, based on their collective experience and their operating policies.

It should be noted that lending agencies such as the EDC in Canada and the Export-Import Bank of the United States have an additional objective. That goal is to facilitate the sales of goods and services by Canadian and United States firms respectively. Most countries have lending agencies with similar objectives.

For international projects that require long-term financing, the choice, by the foreign client, as to which country will provide engineering services will often be determined by which country offers the most attractive arrangement.

Engineering Work Outside Canada Is Different

When considering the prospects of doing international engineering work, either as an individual or on behalf of your employer, you should investigate the rules, regulations and accepted practices at the place of that work because they could differ drastically from ours in Ontario and those in other parts of Canada. While this is just a partial checklist, here are some questions typical of those you will want answered.

In that country:

1) Are practising engineers required by law to hold a licence; and if so, what are the legal requirements and qualifications the engineer must have; and how many engineers are registered in that country?

2) Will you and other foreign engineers be required, by the law of the country, to obtain a licence to practise there; and if so, what are the requirements for obtaining it; how long will it take and how much will it cost?
3) Are foreign consulting engineers required to have local representatives to practise there, and if so, must those representatives be engineers; must they be citizens of that country, or may they be Canadians or other non-nationals who are resident there?
4) Must there be local participation in any contract with a Canadian consulting engineer; and if so, to what extent?
5) What is the official and the commercial language in the country; what system of weights and measures is used?
6) What is the unit of currency; what is the exchange rate and how stable is it; what, if any, restrictions will be imposed on taking salary or fee out of the country?
7) What proportion of take-home pay must be paid by the employer for social costs such as social security, unemployment insurance, health and death benefits, vacation and holiday pay, year-end bonuses, retirement and severance pay?
8) What income tax rules, Canadian and host country, will apply; and what foreign-exchange restrictions will there be?

The answers to these and to the other questions, which of course there will be, will depend on the particular host country. You should do your best to find out what the answers are before making a commitment to do engineering work in a foreign country.

CHAPTER 9

COMMUNICATIONS

The art of expressing ideas by words and/or symbols.

In its broadest sense the scope of communications goes far beyond the written word and includes the many sounds of words and music, histrionics in its many forms, and eyeball to eyeball signals. However, without in any way suggesting that these other ways of communicating a message are not important, or that they are less important, the theme of this chapter is restricted to written communications.

Communications is a critical element in professional engineering practice. In any engineering undertaking, the objective and how to achieve it must be communicated, usually in words. The undertaking could be a firm or a joint venture submitting a proposal to provide the required engineering services on a major project. When such a proposal is accepted, even provisionally, there will be preliminary and feasibility studies; if it clears that hurdle there will be more studies, preparation of designs and specifications. Then will come change orders, inspection reports and eventually the final inspection and acceptance report, probably with a list of deficiencies. This is just one example that shows the importance to engineers of reports and other documents, that is, composing words to transmit thoughts—communications. Another example, while not such a major project in some respects, could be of even greater importance to a young engineer: the preparation of an application for a job.

An excellent article, "The Practical Writer," in *The Royal Bank Letter* (published by the Royal Bank of Canada, Vol. 62, No. 1

Jan./Feb. 1981), is reproduced here, with minor editorial revisions, by kind permission of the Royal Bank of Canada. That permission is gratefully acknowledged by the authors who consider the message in *The Letter* to be of particular interest and of prime importance to practising professional engineers.

The Practical Writer

Written words form the mainstay of communications in organization.
But often they fail to do the job.
Here is a guide to writing that means business.
There's nothing to it but blood, sweat and tears:
then review, revise, check and double-check!

From time to time most business and professional people are called upon to act as writers. You may not think of yourself as such when you dash off a personal note or dictate a memo, but that is what you are. You are practising a difficult and demanding craft, and must meet its inherent challenge. That is, to find the right words and to put them in the right order so that the thoughts they are intended to communicate will be understood.

The purpose of any written message is to transmit the writer's thoughts—on some subject—to the reader's mind. This is so, especially in businesses and institutions where written words carry so much of the load of communications. The written traffic in any well-ordered organization is heavy and varied—letters, memos, reports, policy statements, manuals, sales literature and other messages.

The objective is to use words in a way that serves the organization's aims. Unfortunately, often a written communication does not achieve this objective. Some writing gives rise to confusion, inefficiency, and ill-will. Almost always this is because the intended message does not get through to the receiving end because it was not presented effectively.

An irresistible comparison arises between writing and another craft that most people have to practise sometimes: cooking. In both fields there is a wide range of competence, from the great chefs and authors to the occasional practitioners who must do the job whether they like it or not. In both, care in preparation is essential. Shakespeare wrote that it is an ill cook who doesn't lick his own fingers; it is an ill writer who doesn't work at it hard enough to be reasonably satisfied with the results.

Unlike bachelor cooks, however, casual writers are rarely the sole consumers of their own offerings. Recluse philosophers and schoolgirls keeping diaries are about the only writers whose work is not intended for other eyes. If a piece of writing turns out to be an indigestible half-baked mess, those on the receiving end are usually the ones to suffer. This might be all right in a book, because the reader can toss it aside anytime. But in organizations, where written communications command attention, the recipient of a sloppy writing job must figure out what it means.

The reader is thus put in the position of doing the thinking the writer failed to do. To make others do your work for you is, of course, an uncivil act. In a recent magazine advertisement on the printed word, one of a commendable series published by International Paper Company, novelist Kurt Vonnegut touched on the social aspect of writing: "Why should you examine your writing style with the idea of improving it? Do so as a mark of respect for your readers. If you scribble your thoughts any which way, your readers will surely feel that you care nothing for them."

In the business and professional world, bad writing is not just bad manners, it is bad business practice also. The victim of an incomprehensible letter will be annoyed at best or, at worst, will decide that people who can't say what they mean aren't worth doing business with. Write a sloppy letter and it might rebound on you when the recipient calls for clarification. Where one carefully worded letter would have gotten your message across, you may have to write two or more.

Muddled messages can cause havoc in an organization. Instructions that are misunderstood can set people off in the wrong directions or put them to work in vain. Written policies that are open to misinterpretation can throw sand in the gears of an entire operation. Ill-considered language in communications with employees can torpedo morale.

A Careful Writer Must Be a Careful Thinker

In the early 1950s the British Treasury grew so concerned with the inefficiency resulting from poor writing that it called in a noted man of letters, Sir Ernest Gowers, to work on the problem. As a result Gowers wrote an invaluable book, *The Complete Plain Words*, for the benefit of British civil servants and anyone else who must put English to practical use. (Her Majesty's Stationery Office, London, 1954.)

Gowers took as his touchstone a quotation from Robert Louis Stevenson: "The difficulty is not to write, but to write what you mean, not to affect your reader, but to affect him precisely as you wish." To affect your reader precisely as you wish obviously calls for precision in the handling of language. And to achieve precision in anything takes time.

Gowers suggested that the time spent pursuing precision more than cancels out the time wasted by imprecision. People in administrative jobs might well protest that they were not hired as writers, and that they are busy enough without having to fuss over the niceties of grammar and syntax. The answer is that an important part of their work is to put words on paper. It should be done just as thoroughly and conscientiously as anything else they are paid to do.

No one should be led to believe writing is easy. As great a genius as Dr. Samuel Johnson described composition as "an effort of slow diligence and steady perseverance to which the mind is dragged by necessity or resolution." Writing is hard work because thinking is hard work; the two are inseparable. But there is some compensation for the effort required to write well.

The intellectual discipline required to make thoughts intelligible on paper pays off in clarifying your thoughts in general. When you start writing about a subject, you will often find that your knowledge of it and your thinking about it are inadequate. The question that should be foremost in the writer's mind, "What do I want to say?" will raise the related questions, "What do I really know about this? What do I really think about it?" A careful writer must be a careful thinker—and in the long run careful thinking saves time and trouble for the writer, the reader, and anyone else concerned.

The problem is that many people believe that they have thought out ideas and expressed them competently on paper when they have not. This is because they use nebulous multipurpose words that may mean one thing to them and something quite different to someone else. Gowers gives as an example the verb "involve," which is used variously to mean "entail," "include," "contain," "imply," "implicate," "influence," and so on. "It has . . . developed a vagueness that makes it the delight of those who dislike the effort of searching for the right word," he wrote. "It is consequently much used, generally where some more specific word would be better and sometimes where it is merely superfluous."

The Right Word Will Almost Tell You Where It Should Go

There are many lazy man's words lurking about, threatening to set the writer up beside Humpty Dumpty, who boasted: "When I use a word, it means just what I want it to mean." You should not use ambiguous words and expressions. In his book *How to Be Brief*, Rudolph Flesch gives us the first commandment of practical writing: "Be Specific. Specify, be accurate, give exact details—and forget about fine writing and original style."

Style tends to take care of itself if you select the right words and put them in the most logical order; so, to a large extent, do grammar and syntax. Find the right word, and it will almost tell you where in a sentence it should go. That is not to say that grammar and syntax are not important. Words just scattered on a page would not be comprehensible. The rules of language usage also exert a degree of discipline over your thinking about a subject by forcing you to put your thoughts in logical order. Many grammatical conventions are intended to eliminate ambiguity, so that you don't start out saying one thing and end up saying something else.

Most literate people, however, have an instinctive grasp of grammar and syntax that is adequate for all ordinary purposes. The rules of usage (in English more so than in French) are flexible, changeable, and debatable; new words are invented as the language lives and grows, and a solecism in one generation becomes respectable in the next. So while grammar and syntax have their roles to play in written communications, they must not be followed so blindly that the message is not understood by the reader. Gowers quoted Lord Macaulay with approval on this score: "After all, the first law of writing, that law to which all other laws are subordinate, is this: that the words employed should be such as to convey to the reader the meaning of the writer."

Usually Vocabulary Is the Least of a Writer's Problems

Since words come first, an ample vocabulary is an asset in conveying meaning. Oddly enough, though, people who have difficulty getting their written message across rarely lack the vocabulary required. They know the apt words, but they don't use them. They go in for sonorous but more or less meaningless language instead. People who are perfectly able to express themselves in plain spoken language somehow get the idea that the short, simple words they use when talking are not suitable in writing. Thus where they would say, "We have closed the deal," they will write, "We have finalized the transaction." In writing, they "utilize available non-rail ground-mode transportation resources" instead of "using trucks." They get caught in "prevailing precipitant climatic conditions" instead of in "the rain." They "utilize a manual earth-removal implement" instead of "digging with a shovel." When so many words with so many meanings are being slung about, nobody can be quite sure just what the message is.

The guiding principle for the practical writer should be: always use common words unless more exact words are needed for definition. Surely the reason for this is obvious. It is that if you use words that everybody knows, everybody can understand what you want to say.

A common touch with language has always distinguished great leaders. Winston Churchill comes immediately to mind; as John F. Kennedy said, he "mobilized the English language and sent it into battle." Churchill mobilized the language in more ways than in his inspiring speeches. As prime minister of Great Britain, he was that nation's chief administrator at a time when governmental efficiency was a matter of life and death for the democratic world. In August 1940, while the Battle of Britain was at its peak, Churchill took the time to write a memo about verbiage in inter-departmental correspondence. It read:

> Let us have an end to such phrases as these: "It is also of importance to bear in mind the following considerations . . ." or "Consideration should be given to carrying into effect . . ." Most of these woolly phrases are

> mere padding, which can be left out altogether or replaced by a single word. Let us not shrink from the short expressive word even if it is conversational.

Churchill's own wartime letters and memos, reproduced in his memoirs, are models of effective English. It is interesting to speculate on how much his clarity of expression, and his insistence upon it in others, helped to win the war. He was, of course, a professional writer who had earned a living with his pen since he was in his early twenties. He was a master of the English language and, while it may seem ridiculous to exhort modern white-collar workers to write like Winston Churchill, the principles of writing he used are not hard to grasp.

Churchill was an admirer of Fowler's *A Dictionary of English Usage*, and when he caught his generals mangling the language he would direct them to it. Fowler set five criteria for good writing—that it be direct, simple, brief, vigorous and lucid. Any writer who tries to live up to these is on the right track. By keeping in mind two basic techniques you can go some way towards meeting Fowler's requirements. These are:

Use Verbs in the Active Voice Rather than the Passive

It will make your writing more direct and vigorous. In the active voice you would say, "The carpenter built the house;" in the passive, "The house was built by the carpenter." Though it is not always possible to do so in a sentence, use the active whenever you can.

Prefer the Concrete to the Abstract

A concrete word stands for something tangible or particular; an abstract word is "separated from matter, practice, or particular example." Churchill used concrete terms: "We have not journeyed all this way, across the centuries, across the oceans, across the mountains, across the prairies, because we are made of sugar candy." If he had couched that in the abstract, he might have said: "We have not proved ourselves capable of traversing time spans and geographical phenomena due to a deficiency in fortitude." Again, there are times when abstractions are called for by the context because there are no suitable concrete words, but do not use them unless you must.

Sticking to the concrete will tend to keep you clear of one of the great pitfalls of modern practical writing, the use of "buzz words." These are words and expressions that are used not because they mean anything in particular, but merely because they sound impressive. It is difficult to give examples of them because they have such short lives; the "buzz words" of today are the laughing stock of tomorrow. They are mostly abstract terms ending in -ion, -ance, -osity, -ive, -ize, -al, and -ate, but they sometimes take the form of concrete words that have been sapped of their original meaning. The reason for avoiding them is that their meaning is seldom clear.

Jargon presents a similar pitfall. It has its place as the in-house language of occupational groups, and that is where it should be kept. Do not use it unless you are certain that it means the same to your reader as it does to you.

Combining the active and the concrete will help to make your prose direct, simple, vigorous, and lucid. There is no special technique for making it brief; that is up to you. The first step to conciseness is to scorn the notion that length is a measure of thoroughness. It isn't. Emulate Blaise Pascal, who wrote to a friend: "I have made this letter a little longer than usual because I lack the time to make it shorter."

Use your pen or pencil as a cutting tool. No piece of writing, no matter what its purpose or length, should leave your desk until you have examined it intensely with a view to taking the fat out of it. Strike out anything that does not add directly to your reader's understanding of the subject. While doing this, try to put yourself in his or her shoes.

Be hard on yourself; writing is not called a discipline for nothing. It is tough, wearing, brain-wracking work. But when you finally get it right, you have done a service to others. And, like Shakespeare's cook, you can lick your metaphorical fingers and feel that it was all worthwhile.

You are well advised to heed the creed of the wise writer. There is no such thing as good writing, only good rewriting. (Excerpted from the *Royal Bank Letter*.)

Style in Practical Writing

The word style, used in connection with practical writing, is usually coupled with an adjective. So coupled, it becomes redundant or, at best, is a name for an obvious characteristic of the writing. A plain style means that the writing is plain, a concise style that it is concise, and so on. In contrast to the gross misuses in which style is taken to be the presence of gratuitous embellishment, of vogue words, or of other devices, this common simplistic usage is innocent enough, and is well understood. But it does dismiss the phenomenon of style too easily, and tends to put practical writers off their guard.

In the community of fiction writers one of the accepted doctrines is: "We do not tell readers a story; we provide the stimuli and the readers tell themselves a story."

The pleasurable effect on the reader of such stimuli is achieved by the manipulation of style by a skillful writer. But in our practical writing any such effect is adverse and part of our skill must lie in avoiding it. For in proportion to the incidence of temptation for the readers to "tell themselves the story," the writing departs from the domain of the practical.

Consider a gross example: "Seagoing ships, upbound from the St. Lawrence or down-bound from the Welland, enter Toronto harbour by the Western Gap." At best, we take this to be telling us that "deep-draft vessels use the Western Gap," and we wonder why the author has not simply said so. But it could have been even more obscure had the author put the words in a different sequence, dropped a comma, and written: "Upbound for the St. Lawrence or down-bound from Welland, seagoing ships enter Toronto harbour by the Western Gap." The writer has now beclouded the message by music—a song about ships from the four corners of the seven seas. Some train of thought, established in a preceding context, might have survived the first version, but hardly the second. The first version may have been meant to be practical, but neither is in plain writing.

Now consider this example: "Keep notes in ink in a bound, stitched notebook." You will find it hard to state this rule, using any words you like, and at the same time deliberately work in an extraneous thought, without sacrificing the plainness of the instruction. If you can't do it by trying, you are not likely to do it inadvertently and we arrive at this axiom: "Stick to plain writing, and you need not worry about your readers being diverted from the channel of thought you intend them to follow."

While not a part of the argument for plain writing, the reason this rule is used is related to ethics in engineering work. It is a good rule to follow in any situation where notes might be used later as evidence in some legal action. The precautions of keeping the notes in ink and in using a bound, stitched notebook are to prevent fudging of the notes by erasing and changing them and/or removing a page and replacing it with different notes on the page substituted.

Edmund Gosse, one of the contributors to the *Encyclopaedia Britannica*, (11th edition, 1910-1911), presents many and varied comments and quotations on style in literature. One is that Flaubert's "ingenious definition of style seems to strain language beyond its natural limits"; another is that "all these attempts at epigrammatic definition (of style) tend to show the sense that language ought to be, and even unconsciously is, the mental picture of the man who writes." This can be disconcerting to the unpractised practical writer. Having been assured that if you write plainly you need have no worry about the unwanted effect of style already drawn to your attention, you are now bedevilled by the suspicion that what your typist is reproducing at 80 wpm is a sort of "portrait of the artist" that may cross a hundred desks. How shall you protect yourself against this exposure, or, if you cannot, how can you render yourself more photogenic?

Plain writing offers no refuge. It is the key not to avoiding or disguising the "mental picture of the man who writes," but to the control of it. It is a control that permits the image to reveal the writer only to the extent that the writer relates to the subject written about. If others enter and are represented by the writer (boss, group, firm) what the writer says of or attributes to them must be included in the control so that they are pictured only to the extent that they relate to the matter in hand. But let us not risk "straining language"; here is an example:

"Pay to the Order of ____________ ____________ dollars"

In an age of euphemistic writing, this form has survived because the style precisely fulfils the function intended. Try adding "please," and see the style vanish before your eyes, and the image of the writer (the person who signs the cheque) slips out of focus. As it stands, let us look at what the style says.

> This conventional form was composed, not by the writer who signs it, but by the Bank. We know that, and it is visible in the style, for custom allows the bald command form (drill sergeants and the legal summons excepted) only in writing composed by the party commanded.
>
> Writers who sign the cheque say of themselves (are, in fact, forced to say of themselves when they use the form) that they have the right to command and are willing to assume responsibility for the consequences. They declare themselves, at this moment at least, one with the whole edifice of trust and probity on which such transactions are based.

No more is exposed of the writer than that which relates to the purpose of the writing, although the style transmits a lucid picture. This is the comfort offered any who may be uneasy about self-revelation in practical writing.

Gosse offers this advice: "The errors principally to be avoided in a cultivation of pure style are confusion, obscurity, incorrectness and affectation."

We have mentioned the presence of others in the picture drawn. In Robert Graves' *Claudius the God* the Roman emperor says: "There is an art in playing second fiddle. I know. I have practised it all my life." With us, this is a common position.

Often we write to express decisions that are not our own but have been taken by others. About three generations ago, in such circumstances, writers portrayed themselves by using such formal starts as "I am directed to inform you . . ." They thus declared their part in the business in the first person, and charted the course of their self-portrayal through the whole communication. Using today's less cumbersome formulation, as in writing "Our design group explains the seeming discrepancy by citing the new code requirements," the writer's position of intermediary is implied, although not stated. In both forms, the image pursued is that of a knowledgeable man or woman, able to put forward the arguments with due respect for the dignity of the party addressed and of the writer's principals.

Writing where they are themselves the authority (having presumably carried out the study or inspection that is the subject matter), young engineers must get away from the pupil-teacher style they used, and probably were encouraged to use, in the undergraduate laboratories. It is no longer acceptable to display knowledge just to prove that you have it. For sophisticated readers, elements of classroom recitation reflect only immaturity. Writing on your own authority demands that you construct, for any situation, an answer to the age-old question, "Who am I?"

To the imaginative writer it has been said, "If you would have me weep, you must weep first." For us, if we would add to our practical writing an image of the writer that will illuminate it, as it were, for the reader, we must establish the image in our own minds first, before we select and adopt the norms of expression that will render and sustain it.

The "norms of expression" a writer uses are selected from a memory store, reviewed by a critical faculty, modified, rejected or accepted—all at high speed, all within the brain. Range and variety of the items in store are gained by our reading, and the critical faculty is nurtured by the discernment we apply to the reading. But, considered as a source for augmenting the store or for sharpening our perception, our reading may be random and repetitive. The program, then, consists of a regime—fortunately an easy and entertaining one—of reading books that illustrate "norms of expression," for the most part by discussing hundreds of key words and often setting them in various contexts, showing not so much what is right and wrong, but suggesting what the example is saying to an observant eye and ear about the subject matter and, frequently, about the writer. Three of these are listed below. All are famous, all are written with the gentleness and humour that scholarship seems to achieve in teaching. The listing is in alphabetical order of the authors' surnames, so no favourite is indicated; we recommend all three as an inexpensive little array of bedside companions. Read them and enjoy them!

Fowler, *Modern English Usage* (Oxford).
Gowers, *The Complete Plain Words* (Pelican).
Strunk and White, *The Elements of Style* (Brett-MacMillan).

CHAPTER 10

PROFESSIONALISM – EXPERTISE IS NOT ENOUGH

There are professional engineers' associations in all of the provinces and territories in Canada. Even though there are differences in the wordings, the objectives of the Code of Ethics of each of these associations are substantially the same. The message is that ethical means "conforming to the standards of conduct of the profession or group."

The Definition of Engineering as a Profession

Chapters 1 and 2 of this book, regardless of their titles, do not provide a definitive description of the engineering profession. This much they tell: in Adam Smith's "philosophers" of Chapter 1 and in the "professions by the score" spoken of by Oswald Hall, Chapter 2, we are contemplating "learned professions" (i.e., we dismiss the simpler meaning of the word as vocation and demand the possession of intellectual skills and experience). To go further, we have to explain the modifying word "engineering," and if we clip down the 50-year old definition by the Engineers Council for Professional Development to its bare bones, we get the single-sentence statement:

> Engineering is the profession concerned with utilizing, economically, the materials and forces of nature.

This tells us something about engineering but nothing about what separates the concern of the professional engineer from that of scientists, technologists, skilled tradesmen, etc., whose work is addressed to the same end. So we ask what distinguishes the professional and, whatever it is, can it be identified so as to apply to all professions? And, if it can, should we recognize a "theory of professionalism" which might guide us in assessing the rights and wrongs of conduct?

I believe that such a notion is valid and, within the space allotted me, I will express my conviction. But to argue it and sustain the argument by references to the engineers and academics, mostly wiser than I, whom I have read or with whom I have discussed it, would require many hundreds of pages.

Start, then, by considering that every problem is a human problem, whatever the field in which it lies (as does Oswald Hall in the introductory paragraphs of Chapter 2). Two broad classes of human problems are:

(a) Well-structured or well-defined problems.
(b) Ill-structured or ill-defined problems.

As undergraduates, we used textbooks full (necessarily) of class (a); we were being taught the rationale of logical deduction so often remembered as formulae; it may have taken us some years to get rid of the false idea that simple, universal solutions can be found for all human problems, or we may, on the other hand, have profited early in life by realizing the preponderance of class (b).

Realizing this preponderance and that ill-structured human problems have been with us always (terrorism, poverty, etc.), all multiplying in the complicated systems of civilization, and that ill-structured problems exist for individual human beings, for industries, for corporations, governments, and so on, I contend that the expertise needed for dealing with all ill-structured problems is exactly that of the professional, and I propose three basic principles for defining this expertise and its application:

I. A professional is a person who has had sufficient experience and has sufficient knowledge to be able to deal with the problems of humans in complex systems of human activity.

The second principle is based on how professionals are observed to operate; it was expressed to me by Dr. Peter E. Korda, a professional engineer of Columbus, Ohio, and I think of it as "Korda's Principle."

II. A professional is a person who, with the cooperation of the client, has special competence for being able to acquire more knowledge of the personal or business affairs of the client, in a particular field, than the client himself possesses.

We recognize Korda's principle in patient-doctor or client-lawyer contacts. It may be novel to think of it as applying to client (or employer)-professional engineer relationships, but it plays a bright light on implicit mutual obligations, such as respect and trust.

The third principle of professionalism is basically due to Dr. Rollo May, a psychiatrist who finds that the courage to create is a characteristic of professionals in all fields and has written a justification of this opinion. (May, *The Courage to Create*, Basic Books, Inc., New York, 1975)

III. A professional is expected to make a deep and personal commitment to each client as to the reliability and quality of services provided and to realize the small but real possibility of error.

These three principles emphasize the personal nature of the work of a true professional as compared with that of one who is merely an expert, no matter how recondite the latter's field. Happy is the client whose business and personal relationships are all blessed with mutual respect; indeed, few men, women, or corporations achieve such a state. But the client who retains or employs a professional without being confident that mutual trust will exist, without willingness that the professional may come to know more about the activities in the field involved than he knows himself, is in reality buying no more than the services of an expert.

Implicit, then, in the duty of the professional is his store of knowledge and, by virtue of it, authority beyond that knowledge and authority of the employer or client in some part of the latter's affairs. The duty of confidentiality is inescapable. But, a member of a profession that is duly recognized, has been granted privileges, and licensed by public authority so that it may operate in the public interest, shares the profession's obligation to safeguard and further that interest. What if the two obligations, the personal to the client or employer and the other to the public, are in conflict?

On Serving Two Masters

Whistleblowing is the name attached to what children call telling tattletales. We hear of it in industrial situations, where one worker betrays another, or one department another, by calling managerial attention to fudged workmanship, thievery, or sabotage. For a professional to betray a secret of his employer's to which he is privy by virtue of his duties is a grave act of departure from his personal obligation, a breach of confidentiality, a rupture of mutual trust. Yet the same trivial name, whistleblowing, where a secret is made known to the world at large, is used for it.

A fictional instance of a conflict between the duty of confidentiality owed the employer by a professional engineer and the engineering profession's duty to protect the interest of the public, which has granted it its special status, is given in Chapter 3, under the heading "The Code of Ethics," case (iv). Note that the engineer does not divulge his information to the public. He would have no more dreamed of providing grist to the mill of the media as he would have of continuing in his job after his private communication to the Standards Association, even had there been protection of that job under law such as has been proposed in some localities. Proposed, I must add, by naïve believers that such complex human situations as that portrayed can be tucked away under a legislative rug.

A Definition of Engineering

I offer, finally, for use as a privately held (I think it a little beyond easy understanding for others than engineers themselves) definition of the profession:

> Engineering, *n.* — the profession that deals with the ill-structured problems human beings encounter when involved with the complicated systems associated with their physical environment (the materials and forces of nature).

CHAPTER 11

QUESTIONS

This chapter contains 40 questions typical of those used on the examination paper in professional practice and ethics at the APEO examinations during the past several years. For the examination each candidate is given a sheet which contains two sections of O.Reg. 538/84: s. 86, Professional Misconduct and s. 91, Code of Ethics.

No textbooks or other aids are permitted. If you have any doubt about the meaning of any question, give a clear statement of how you have interpreted it. Each jurisdiction has its own rules for the conduct of its examinations and so you should find out the rules that will be used in your case.

The nature of the subject is such that it is not enough to give a simple yes or no answer to the question, "Was it unethical?" for the engineer to do whatever it was the question states that he or she did. The candidate is expected to appreciate the ethical considerations involved and to explain how the yes or no answer was arrived at. Also, because of the complexity of a given situation, it may not be an open-and-shut case. Often the answer should be such as, "Yes, but we should also consider this way of looking at it," with some discussion from that point of view. The mark the examiner assigns will depend not only on the analysis and reasoning presented by the candidate but on how the answer is reported.

Wherever appropriate in the context, the singular is intended to include the plural and the masculine to include the feminine.

1. A technical paper based on extensive research was selected by an engineering society as the best in its field. Awards were made to two of the three co-authors of the paper at a national meeting of the engineering society, but the senior author was not given an award because of a

rule of the society that restricts its awards to those who are members at the time a paper is submitted. The non-member co-author was not given credit at the award presentation.

(a) Was it unethical for the society not to give an award to one of the co-authors?

(b) Did the society officers and award recipients have any obligation to give credit to the non-member co-author?

2. An owner retained engineer Smith for certain engineering services under written agreement but before Smith had completed the work the owner notified him that the agreement was being terminated. Subsequently the owner retained engineer Jones to do the same work. Smith informed Jones, in writing, that termination of his agreement had been by unilateral action of the owner and that he, engineer Smith, had not accepted it. Was it unethical for Jones to undertake the work in view of the unilateral termination of Smith's agreement, without his consent and in spite of his objections?

3. John Doe, an engineer employed by a testing laboratory, represents his firm on a standards committee for automobile products. All but two of the 10 members of this committee are engineers. After much deliberation the committee arrived at a consensus but Mr.Doe was violently opposed to the end result and registered his objection. After careful consideration of this objection the committee passed the standard for formal publication. Subsequently the laboratory received a contract to test automotive products to the standard and Mr. Doe was assigned the job of supervising the tests and compiling the final report which indicates that the samples meet the requirements of the standard, and signing the report on behalf of his firm. He objects because he considers that his signature on a report that states that the product conforms with the standard would indicate that he endorses the standard. In this case he does not consider the standard to be proper.

   Is he correct in his assumption? What action should he take?

4. An engineer was retained by a government department to make a study of computer programming methods and

techniques for the economical extraction of certain metals from ore. He undertook extensive investigations, conducted experiments and submitted a comprehensive report which contained detailed recommendations to solve the problems set out in his assignment. He was paid an appropriate fee for his services and the project was terminated. The government published the report and made it available to the public. Two years later the engineer was approached by a commercial mining company that had a problem similar in scope and content to that which he had studied under his contract with the government. He was requested by the mining company to act as its consultant and to recommend methods to improve its operation in this type of work.

Does he have an ethical obligation to inform the mining company, before he accepts the assignment, of the existence and availability of the report he had prepared for the government?

5. For many years the XL Company has manufactured a product that enjoys a high rating in the industry and with the public. Competing manufacturers have now introduced a similar product of lower quality at lower cost, and this competition has caused a serious decline in the sales of the product manufactured by XL. To meet this competition XL instructs its engineers to redesign its product so that it may be made available to the market at a lower price. Upon receiving these instructions some of the engineers question whether such an action would be consistent with the Code of Ethics, because a lower-quality product under the same brand name would mislead the public into accepting a product of lesser quality in the mistaken belief that it meets the high quality standards with which the product has been associated in the public mind for many years.

   Do the engineers have a proper interest and ethical obligation to protest the company's decision, or to refuse to redesign the product?

6. An engineer enters into a contract with a public body whereby he agrees to conduct such field investigations and studies as may be necessary for a determination of the most economical and proper method for the design and construction of a water-supply system; he is to pre-

pare an engineering report, including an estimate of the cost of the project, and to estimate the amount of bond issue required.

The contract provides that, if the bond issue passes, the engineer will be paid to prepare plans and specifications and supervise the construction, and will be paid a fee for his preliminary services. If the bond issue should fail, the public body would not be obligated to pay for the preliminary work. The public body is prohibited by law from committing funds for the preliminary work until the bond issue is approved.

May an engineer ethically accept a contingent contract under these conditions?

7. Engineers of Company A prepared plans and specifications for machinery to be used in a manufacturing process and Company A turned them over to Company B for production. The engineers of Company B, in reviewing the plans and specifications, came to the conclusion that they included certain miscalculations and technical deficiencies of a nature that the final product might be unsuitable for the purposes of the ultimate users, and that the equipment, if built according to the original plans and specifications, might endanger the lives of persons close to it. The engineers of Company B called the matter to the attention of appropriate officials of their employer who, in turn, advised Company A of the concern expressed by the engineers of Company B. Company A replied that its engineers felt that the design and specifications for the equipment were adequate and safe and that Company B should proceed to build the equipment as designed and specified. The officials of Company B instructed its engineers to proceed with the work.

   Under these circumstances what should the engineers of Company B do now?

8. An engineer working with a gratuitous service organization with the objective of "assisting those working to raise living standards in developing areas of the world" has inquired as to the extent to which he may ethically provide engineering assistance through this volunteer organization. A number of inquiries for technical assistance are from overseas field workers associated with var-

ious volunteer organizations who are seeking technical advice related to problems in the areas of the world in which they are stationed.

A typical request for assistance directed to the engineer was for a method of improving a primitive African water supply, which consisted of uncovered holes intercepting ground water. Those requesting such technical advice are not in a position to pay for it.

Is it ethical for an engineer to provide technical advice, for no fee, to persons or organizations through a volunteer organization?

9. An owner retained an architect to prepare plans and specifications for a building, using the standard CCDC 2 contract form. The architect retained a structural engineer for the structural portion of the plans and specifications, and the building was erected. Both professionals completed their respective portions of the contract, except the execution of the required certificate of compliance. During the progress of the work, the owner made progress payments to the architect, and the architect paid the appropriate amount from these payments to the structural engineer. However, when the building was completed and ready for occupancy, the owner still owed and refused to pay the architect a substantial sum that was due under the contract, and so the architect owed the structural engineer a proportionate amount. The owner alleged that there had been several deficiencies in the work of the architect and refused to pay him the balance due. The owner requested the city officials to issue to him an occupancy permit, and they requested the architect, who in turn requested the structural engineer, to certify that the structural system was completed in compliance with the applicable building code and regulations. Such a certification is required before the city may issue an occupancy permit to the owner. The structural engineer refused to provide the certification until he had been paid for his services.

   Is it unethical for the engineer to refuse to provide the certification that would enable the owner to secure the occupancy permit, because he has not been paid for his services?

10. An engineering firm with extensive experience in the

design of industrial equipment and processes proposes to establish a value-engineering division, under which it would offer to clients a value-engineering service. The plan of operation contemplates offering to clients in the industrial field a review of their designs, equipment, products, processes, and so on. Following such review, the firm would suggest to the client revisions in the design or production procedures that would produce the same result at less cost to the client without sacrifice of quality or safety. If such savings are realized, the client would pay the firm a portion of the savings, determined by negotiation before the service is rendered, in accordance with the extent of work involved. If the client determines that the proposed changes are not feasible or desirable for any reason and does not utilize the suggested changes it is not obligated to pay the firm for these services.

Is this proposed method of operation ethical?

11. Engineer A is employed by an industrial corporation. His immediate supervisor, Engineer B, is chairman of a civic committee responsible for retaining an architect to design a civic facility. When Engineer B received the completed plans and specifications from the architect, he directed Engineer A to review them in order to (1) gain knowledge, (2) suggest improvements, (3) assure their compliance with the specified requirements.
    (a) By so directing A, is B guilty of either professional misconduct or of a violation of the Code of Ethics?
    (b) Will it be unethical for Engineer A to carry out the instructions given him by Engineer B?
12. An injured worker is involved in a proceeding before a worker's compensation board relative to the amount of compensation to which he is entitled. The determination rests in large measure upon the conclusion of the board as to certain technical details related to the accident. The worker asks an engineer to appear before the board as an expert witness, but states that he is indigent and cannot afford to pay the engineer for his services.

    The engineer is willing to assist the worker, but asks whether he may ethically do so: (a) on a contingent-fee arrangement, whereby he would be paid a percentage of the amount received by the worker, or (b) as a compassionate and gratuitous action.

Is it ethical for an engineer to provide services as an expert witness for an indigent client on either a contingent or free basis?

13. A company that regularly undertakes projects for the federal government employs several registered professional engineers. During slack periods this company "stockpiles" its engineering staff to the extent necessary so as to have them available when needed for future work and assigns them to clerical and other sub-professional duties. During these assignments the direct supervisors of the engineer involved are not engineers.

    Is it a violation of the Code of Ethics for the engineers to serve in clerical and sub-professional capacities during slack periods of their employer's operations?

14. A manufacturing company has contracted to develop and produce a completely automated mass-transporation system. Public safety would be endangered by a failure of the system, if one were to occur. A series of engineering tests were carried out on the various components during the development period, but one major assembly did not perform satisfactorily. The engineer who is manager of the department responsible for the project reported the failure to his superiors. He was told, however, that in order to meet the contract commitments the equipment would be shipped to the client without informing the client of this failure. The engineer objected to this decision and learned subsequently that shipment to the client had been made.

    What, if any, further action should the engineer take under these circumstances?

15. An employee, X, worked for Engineer A for five years and had progressed from draftsman to designer. A expected that, with further training and experience, X would continue to progress. Engineer B, in the same city, approached X directly and offered him a job with B's firm without notice to, or discussion with, A. X accepted the offer and, after giving two weeks' notice to his employer, A, left to work with B.

    Did B act unethically?

16. An engineer is employed by a large consulting engineering firm and his work includes the designing and specifying of electrical equipment. He owns shares in a large,

well-known and reputable electrical manufacturing company; his share holdings amount to less than 1% of the company's outstanding shares.

Is it a violation of the Code of Ethics when, in the course of his work, this engineer selects and specifies equipment made by the company in which he holds stock?

17. A member of the city council is chairman of the council's finance committee, which deals with and makes recommendations regarding appropriations for projects undertaken by the city. One such project is a pollution abatement project, for which funds were allocated. The chairman of the council's finance committee is one of the principals in a consulting engineering firm that has established a good reputation in the pollution control field, and the firm submitted a proposal to the council to provide the engineering services required for the project under consideration.

    Under these circumstances, was it ethical for this consulting engineering firm to offer to undertake this engineering work?

18. An engineer employed by Company A is assigned by his supervisor to develop processing equipment for the manufacture of certain chemical products. In his previous employment with Company B, the engineer had participated in the development of similar equipment. The technical information concerning the equipment has not been published in the technical press, or otherwise released. Because of his earlier experience, the engineer is familiar with the equipment and the principles of its design. His superiors in Company A suggest that this knowledge will be useful in developing similar equipment for their use and expect him to make his knowledge concerning the particular equipment available to Company A to aid in the development of the similar equipment.

    Would it be ethical for the engineer to apply his knowledge to the development of equipment for his present employer based on experience and information gained in similar work for a previous employer, without the consent of the previous employer?

19. Engineer X is one of the partners in a consulting engineering firm, XYZ, whose services include designing,

specifying and supervising the installation of mechanical-electrical systems in buildings and factories. X proposes to form a company that would contract to provide system operation and maintenance services for such facilities. He would be the major shareholder in this company and expects that most of the contracts the company would get would be for work on installations that had been engineered by XYZ.

Would the operation of the proposed company, in the manner envisaged, be consistent with the requirements of the Code of Ethics? Comment on the various points involved.

20. Engineer A learned that an engineering position with the CBC was vacant and he applied for the job. After the interview he was told by the interviewer that he was being considered favourably for the appointment, but he was not hired nor was any commitment made to him. Engineer B learned of the opening and, while knowing that A was being considered favourably for the appointment, applied for the same position. In presenting his own qualifications, B made no reference whatever to A. When A learned that B had applied for the position, while knowing that he was being considered favourably for it, he complained to his professional engineering association, alleging that this action by B was unethical.

    Was B's action unethical?

21. Most of the duties of an employee at a consulting engineering firm are related to finding and exploring prospects for engineering service contracts for the firm. This employee is not a professional engineer; he is paid a salary plus a commission on the contracts he arranges for the firm.
    (a) Is it ethical for an engineering firm to have a non-engineer as a representative to discuss engineering aspects of a project and contract negotiations with a client?
    (b) Is it unethical for a firm to employ a non-engineer as a representative to solicit contracts for its services?
    (c) Is it unethical to compensate such a representative on a commission basis?

22. An engineer becomes aware that one of the products his company produces does not meet the standards required

by law, and when he brings this to the attention of his immediate superior, he is either overruled or ignored.

What action should the engineer take? Does it make any difference if the safety of the public is not involved?

23. A building contractor engaged a professional engineer to design and prepare drawings for the formwork and scaffolding for a reinforced-concrete building, to meet the requirements of construction-safety legislation. The engineer did this, and affixed his seal and signature to the original tracings, which he turned over to the contractor. Was this acceptable professional practice?

    Later the contractor engaged the engineer to inspect the scaffolding as built. He found that in many significant parts his design had been ignored, and the contractor's superintendent had built it the way he thought it should be built.

    What should the engineer do? Discuss this situation and the engineer's professional responsibility for the safety of the workers.

24. A company has a large, integrated operation that includes consulting engineering services, construction, and manufacturing. Its brochure and advertisements state that it is engaged in all of these activities and that it will provide all or any portion of these services, as may be the wish of a client. The company usually acts for clients on major projects such as dams, power plants or factories, and has a separate division for each of the main functions. All of its engineering operations are directed by registered professional engineers.

    Because of its other activities, is the practice of consulting engineering by this company a violation of the Code of Ethics?

25. A Canadian professional engineer is working in a foreign country for a client building a power station. He is acting as technical adviser to the client. The client is directly supervising all construction labour.

    The client does not have any apparent safety procedures for his labour; no hard hats, no safety shoes, in some cases no shoes. Holes in floors do not have safety barricades. The conditions are clearly such that they would be unacceptable in Canada. Even assuming the poor safety conditions will not affect the technical aspects of the power station, clearly they affect the safety of workers.

Would it be ethical for the Canadian engineer not to take any action? What kind of action could he take? Do you consider that it is likely that the poor safety practices could affect only the safety of workers and not have any relation to the technical aspects of the power station?

26. The engineering staff of a municipality prepared designs and specifications for the paving of a six-mile stretch of road. Two designs were prepared, one for concrete pavement and the other for asphalt, with the understanding that the two would be of equal quality and would give equal service. Tenders for the concrete pavement averaged about 15% higher than those for asphalt. All tenders were rejected and the engineers were instructed by the chief engineer to revise the concrete design to a lower standard, but not to change the asphalt design. In the second set of tenders the bids on the revised design for concrete pavement were about 7% higher than the new asphalt bids. The chief engineer recommended that the contract be let to the low bidder on the revised concrete design.
    (a) Was it ethical for the chief engineer to recommend the award of a contract based on a design to a lower standard, at a higher price?
    (b) Was it ethical for the chief engineer to have the design revised to a lower standard in order to reduce the price of concrete but not asphalt?
27. A large multinational corporation selected and interviewed three consulting engineering firms in Canada regarding the engineering services it would require in connection with the designing of, and the supervision of the construction of, a new plant the corporation was planning to have built.

    Each firm, in its proposal, stated that the fee would be that recommended by its professional engineering association for complete engineering services on a project such as the one described.

    Later, the corporation asked each firm to state the amount by which it would reduce its fee if the corporation provided the following portions of the overall engineering services:

(a) preliminary engineering studies and a report that contains a suggested layout for the plant;
(b) all field engineers and inspectors required to supervise the construction of the plant.

The three firms got together and discussed this request and agreed on the amount (the same figure for each of the three) by which they would reduce the overall fee to allow for the data and field staff to be provided by the corporation.

Was it unethical for these three firms to confer and agree on an amount to allow for these data and field staff?

Would it have been competitive bidding for each of the firms to determine independently an amount by which it would reduce the fee?

28. A young Canadian graduate in chemical engineering is employed by the Canadian branch of a large multinational chemical company. The Canadian branch manufactures, among other things, a wide variety of solvents, paints and other types of finishes, using formulas developed in the research laboratory located at the American head office of the company. A Canadian manufacturing company is a principal customer. A Ministry of Labour inspector has required the company to provide the ministry with information about all the ingredients used in the products turned out by the plant, particularly regarding their possible risk to the health of the factory workers. The professional engineer knows that a key ingredient in one of their products has been linked to a large number of cancer cases among the employees of an American customer. He has been ordered to prepare a report to meet the requirements of the Ministry of Labour inspector.

What course of action should he take?

29. Basically, the Code of Ethics sets the standard of conduct for the professional engineer, yet often the employers of engineers are not familiar with the Code of Ethics or even aware of its existence.
(a) Why then is the Code important to industry?
(b) What could be the consequences to a company of

using persons who are not registered professional engineers in engineering positions?

(c) Is the Code of Ethics compatible with the goals of industry?

30. An engineer in private practice was retained by a client to design and to supervise the construction of a warehouse. Some time later he was asked by another client to provide professional engineering services for a warehouse almost the same as that previously designed by him, except for those minor changes that were required to conform the building with the site. This client suggested that the fee be lower than that charged for the original design services because the engineer could use his same design with only minor changes.

    For this re-use of his design, would it be ethical for the engineer to charge a fee substantially less than he had charged the first client? What are your reasons for this opinion?

31. An engineer who is an expert in his field of technology is often called upon to serve as technical adviser on behalf of parties in lawsuits. His assignment is to provide the attorney for the party that retained him with expert analysis and advice on the technical reasons for the failure that led to the damage. He may also be called upon to testify as an expert witness in support of his findings on the technical aspects of the case.

    The usual arrangement is that he is paid for these services on a per diem basis. For this case he has been requested to provide similar services for a plaintiff on the basis of being paid a percentage of the amount recovered by the plaintiff. If the judgment is in favour of the defendant he will not be paid for his services. The attorney for the plaintiff is handling the case on a similar arrangement for his fee.

    Would it be ethical for an engineer to act as a technical advisor, to serve as an expert witness in a lawsuit on a contingent fee basis?

32. After having been employed by consulting engineer B for several years, engineer A terminated his work with B and started his own practice of consulting engineering. Later

B learned that some of his sub-professional employees were doing work for A on their own time. B is of the opinion that the outside work by his sub-professional employees is so extensive that it adversely affects their productivity while they are working for him.

Did A act unethically by employing the sub-professional employees of B under the conditions stated?

33. A strike by its production and maintenance workers seriously disrupted the operations of an oil company, including the operation of one of its refineries. During the disruption, officers of the company assigned some of the company's engineers to duties normally carried out by the workers who were on strike in order to continue the operation of the refinery on a reduced basis.

    Was it a violation of the Code of Ethics for these engineers to perform these duties?

34. For several years an extensive and costly flood control and hydroelectric project in Canada has been considered by a provincial government. There is no generally accepted opinion as to which of two different design approaches should be adopted. One of these would use one high dam while the other would use a series of low dams.

    A meeting was held by a committee of the provincial legislature to hear and consider comments and recommendations regarding these two proposed design approaches.

    A professional engineer representing the provincial power commission reported that studies he and his professional colleagues had made indicate that from an engineering standpoint the more efficient solution is the one with a series of low dams.

    Another professional engineer representing a private power company reported that his engineering analysis indicates a more effective and less expensive solution is obtained using one high dam.

    Each presented voluminous engineering data to support his conclusion and openly disagreed with the analysis and recommendations of the other.

    Discuss this situation and whether there was a violation of the Code of Ethics by one or the other engineer by criticizing the work and the statements of the other engineer in a public meeting.

35. A municipality owns an arena and was ordered by a provincial government inspector to have the structural adequacy of its arena roof inspected and evaluated by a professional engineer. It engaged engineer A and his report stated that the roof trusses were not strong enough to support safely the snow and wind loading stipulated for that location by the then current Ontario Building Code. Engineer A was then asked to prepare a design to strengthen the roof trusses and to estimate the cost, which he did.

    The municipality learned from an adjacent municipality that engineer B, whom it had engaged to investigate a similar building, had proposed a much simpler and cheaper method of repair and so it decided to consult engineer B, and engaged him.

    Comment on the ethics of the behaviour of engineers A and B and on the responsibilities of the two municipalities to provide for the safety of the people who will use the arenas.

36. In your province or territory there is an Act and a Code of Ethics to govern the members and licensees of the association of professional engineers. How do the requirements of the Act differ from those of the Code of Ethics?

37. You are a professional engineer with XYZ Consulting Engineers. You know that your firm sub-contracts nearly all of the work associated with the set-up, printing and publishing of reports, including art work and editing. Your wife had some training along this line and now that your children are at school she has some free time on her hands. You decide to form a company to enter this line of business together with your wife, your neighbour and his wife. Your wife will be the president, using her maiden name, and you and your neighbours will be directors.

    You see an opportunity to get work from your firm, XYZ, and expect that there will be similar work from other consulting firms. You know about the existing competition and what they charge for their services, so you feel that this would be a profitable sideline business.

Will it be ethical for you to do this? If so, what steps must you take?

38. You work and live in a small community. You carefully selected Brownsville to be your home because you wanted a quiet lifestyle for your family. Your firm is approached by Disneyworld to undertake a socio-economic and environmental impact study in connection with the construction of a world-class entertainment park on prime farm land within your community.

    You visualize the green meadows turned into parking lots, and rush-hour traffic jams during the weekends. You see your quiet weekends destroyed by the blare of carnival music and you see park visitors littering the countryside. And, apart from the change in your surroundings, you can't help asking yourself why you must aid and abet the intrusion of typically "American" culture. There is so much in Canada that is American and you feel strongly a national Canadian identity should be protected at all costs.

    However, your firm accepts the project and you are asked to become the project manager. Can you?

39. The chief of the Indian band on a reservation plans to develop a resort to provide accommodation for vacationers. This resort would be close to the intake of the pipeline which carries water to supply a large city in western Canada.

    Because of the possibility that such a development would cause unacceptable pollution of the water at the intake of its pipeline, the city's water commission intends to take whatever action is necessary to prevent the development proposed by the Indian band.

    (a) What ethical considerations are involved in this situation?
    (b) If you had the authority, how would you resolve this conflict of rights?

40. As chief engineer of the XYZ Company you interviewed and subsequently hired Mr. A for an engineering position on your staff. During the interview Mr. A spoke of his engineering experience in Quebec, where he had worked most recently, and stated that he was "also a member of the Order of Engineers of Québec." You assumed that he

was a licenced professional engineer in Ontario. You had business cards prepared for his use describing Mr. A as a professional engineer, and he accepted and used these cards without comment. Some months later you received a client complaint about Mr. A calling himself a professional engineer when, in fact, he did not hold an Ontario licence to practise. Upon investigation you found this to be true and you terminated Mr. A's employment with XYZ immediately.

(a) Was your action ethical?

(b) Was Mr. A's action unethical?

CHAPTER 12

SOME ADVICE TO EXAMINATION CANDIDATES AND ANSWERS TO SOME QUESTIONS

We recognize that, when writing an examination, your prime objective is to obtain a "pass" standing. The advice we offer here includes that objective and goes beyond it.

The essence of this message is: do your best!

You should study the introductory remarks in Chapter 10.

At the start of the examination, what you should do first is read and make sure you understand the notes at the beginning of the examination paper. Then you should read through all of the paper to find out what, if any, options you have. You will find that to have been a wise investment of your time.

Probably there will be some selection. Choose the questions you feel you would do better on, and the sequence in which to answer them. You will find that time spent on this process at this stage will be time well spent. Based on your knowledge of your speed at writing examinations you should know whether the time available is likely to be a significant factor. If there is some probability that you will not have ample time for composing and writing your answers to the questions selected, you should establish a target schedule and hold to that schedule within reasonable tolerances. Do not get carried away by your special interest in one of the questions and spend so much time on it that you will have to skimp on the others.

Because of the ambiguity of so many words and expressions in the English language the intent of a statement or question is too often misinterpreted by the reader or listener. Not long ago in Canada, an air-traffic controller ordered a snowplow operator to "clear the runway." He intended his instruction to be "get your equipment and yourself out of the way and off the runway because there is an airplane coming in to land." Instead the operator interpreted the message to mean "clear the snow off the runway." The plane hit the plow and forty-three people were killed.

As stated in the introduction to Chapter 10, if you have any doubt as to the meaning of any question, you should give a clear statement of how you have interpreted the question.

Answer the question asked. That may seem to be a redundant suggestion but all too often a candidate has an interest in a point that is only slightly relevant to the situation described in the question and goes off on a tangent. To illustrate this point refer to Question 34. It may be interesting to you to speculate on what technical analyses and conclusions may have been advanced by either or both engineers, and which of the two proposals would be selected by a competent engineer after a thorough, unbiassed engineering and financial study of the proposed project. These are interesting matters to contemplate but neither is what the candidate is asked to do in the final paragraph of Question 34.

The Statutes of Ontario 1984 include the Professional Engineers Act and Regulation 538/84 under that Act. The Regulation includes not only the Code of Ethics but also a section which deals with professional misconduct. Many actions previously considered to be unethical are now classified as professional misconduct.

As stated in the preface of this book, the engineering profession in Canada—with some variations in details and mechanisms—is self-regulating under the Acts which, along with certain by-laws and regulations, are intended to ensure orderly and systematic conduct in relations between our profession and the public.

To become a worthy member of the engineering profession you must develop many talents. Of the many requirements, two of the more important characteristics are competence and honesty. Together they constitute the foundation on which the profession is based.

In this chapter the authors present what they consider to be acceptable answers to some of the questions in Chapter 10. The introductory comments to that chapter are relevant to what follows here.

Along with the examination paper for the APEO Professional Practice Examination each candidate is given a sheet, which contains Section 86 (Professional Misconduct) and Section 91 (Code of Ethics, from Ontario Regulation 538/84). No other books or aids are permitted.

A candidate preparing to write the examination for admission to one of the other professional engineering associations in Canada should obtain the rules and regulations that will apply to the examination he will write.

Most of the questions on the examination paper will be related to a situation that will require your analysis of, and deliberation regarding, the ethical aspects of one or more of the professional-engineering practice features of that situation. The case-studies approach is usually an effective way to carry out such an exercise. Here is an example:

A CASE STUDY

A Canadian professional engineer is working in a foreign country for a client who is building a power station. The client supervises directly all of the construction labour. The Canadian is technical adviser to the client.

The client does not have any apparent safety procedures for his labour: no hards hats, no safety shoes, in some case no shoes. Holes in floors do not have safety barricades. The conditions are clearly such that they would be unacceptable in Canada.

Question 1. While assuming the poor safety conditions will have no effect on the technical aspects of the power station, clearly they can affect the safety of the workers. Is it ethical for the Canadian engineer not to take any action? What kind of action can he take?

Question 2. Do you consider that it is likely that the poor safety practices would affect only the safety of the workers and not have any effect on the technical aspects of the power station?

This "case study" differs in minor detail only from Question 25 in Chapter 10 of this book and is typical of the world outside of the classroom. It separates students and practising engineers into two distinct groups. One group consists of those who think there are absolute answers and seek to find them, often by methods that are little better than guessing. Those in the other group take a systematic approach to assess the information presented in the case. It is important to use a systematic method; achieving an answer is almost incidental. By using systematic procedures you will always develop a reasonable answer.

Notice that the result is described as reasonable—there may be more than one answer. This possibility of having more than one answer seems to surprise many students. It should not be a surprise—life's like that. Just as there may be more than one way to go to work, more than one strategy to achieve an objective, more than one way to pay for a purchase, often there can be several solutions to a problem.

The following notes outline one method of looking at a case systematically. There are five parts to the method:

> identify the problem; identify the question; list the facts; list postulates; interrelate the pertinent facts and useful postulates.

Identify the Problem

The first two parts to this method might appear to mean the same thing but the "problem" and the "question" refer to quite different aspects of the case. One way to view these two aspects, the problem and the question, is to think of them as: a) the problem is what you are given; b) the question requires you to give something back—a reasonable answer.

Another way to think of these aspects is that if there was not a problem, there would not be any questions; the problem leads to the questions.

In looking back at the case, the first paragraph describes an engineer working on a foreign project that involves a power station. This paragraph provides some information but does not indicate the existence of a problem.

The second paragraph describes some safety-related matters. Indeed, there are serious construction-safety deficiencies reported. This clearly seems to be a problem. Hence the first step of the method seems to be satisfied by identifying a problem of construction safety.

Can there be more than one problem in a case? The answer is yes. However, having found at least one problem it is not forbidden to move on to the second step of the method.

Identify the Questions

In this case two questions appear to be clearly identified and labelled as Question 1 and Question 2.

Cases can be presented without highlighting the question so clearly, nor do questions have to be at the end of the case, as in this example.

Moreover: is it true that there are only two questions? Under Question 1 there are really two questions and these are fairly obvious, as they are in separate sentences, both ending in question marks. Thus in total there seem to be three questions in this case.

List the Facts

The facts are things or information given in the case that can be considered to be true. In this case the list includes:

> It is a foreign project.
> There is a client.
> The client supervises the labour force.
> The engineer is a technical adviser to the client.
> The engineer is a Canadian registered P.Eng.
> The project is a power station.
> There are no apparent safety procedures: no hard hats; no safety boots; no safety barricades.
> These working conditions would be unacceptable in Canada.
> The Canadian engineer knows the conditions are unsafe.

Having listed all the facts that seem to be given in the case the next step is to list postulates.

List Postulates

Listing postulates requires some use of imagination. In fact, in the first instance it sometimes can be a useful exercise to put down all the possibilities that come to mind, no matter how fanciful or unlikely they may be. It is always possible to cut down a long list by eliminating unreasonable or inappropriate items afterwards.

What are the postulates? Postulates are hypotheses—any things, conditions, features, attributes, knowledge, actions or other features that might be possible and that you can imagine might be true in the context of the case but that were not given in the case.

In this instance the following postulates are included, and, of course, there may be more.

> There are safety procedures but the engineer has not seen them.

Safety of labour is not important in this country.
The client does not understand safety practices.
There are safety laws but the client is ignoring them—the laws are not enforced.
There are safety laws and these are usually enforced—there is some other reason the client is not following safety practices on this job.
Poor safety practices will not affect the technical aspects.
Poor safety practices will affect the technical aspects.
The first question is a matter of professional ethics as implied by the wording of the question.
Conversely, the first question has nothing to do with professional ethics but may be pertinent to social ethics.
Canadian ethics, professional or social, do not apply in the foreign country.
The Canadian engineer can offer safety training instruction or guidance.
The Canadian can seek advice from or can notify safety agencies in the foreign country.
There are engineers on the project from countries other than Canada.
It is not a matter of ethics but of commercial interest. Even if it is not a matter of ethics, this would not be a reason to not take some action.
In the first question, ethics is not really an important consideration if at all a valid part of the question.
The Canadian engineer has no authority to take any action.
The Canadian can exercise persuasion in some form to influence the client.
The Canadian is himself endangered by the poor safety practices.

The list of postulates can be a long one. The important result, however, is that some of the postulates will almost always begin to lead to possible answers to the question.

Interrelating Facts, Postulates and Questions

This last step can be considered as a time to discuss facts and postulates that appear to relate properly to the problem. One way to start this is by eliminating ideas that do not seem to be pertinent. For example, it probably does not matter to the problem of safety that the project is a power station. It could have been a chemical plant or an oil refinery with the same kind of problem.

Another way is to consider facts and postulates that tie together. The postulate that not only the local workers but the Canadian also may be endangered by poor safety practices indicates there is a strong incentive for the Canadian to take some action even though it is not a matter of ethics. The argument that poor labour safety practices will affect the technical aspects might help convince the client that safety is important. Volunteering to provide a training program in safety practices is a specific action that can be taken. Other possible actions include seeking help from local government agencies and cooperating with engineers from other countries working on the site to encourage safety practices. Setting a proper example is a form of persuasion.

The above discussion is not the only line of reasoning that may be applied in this case study. A second theme could be developed: since the Canadian is a consultant, among the duties for which he has been retained is to institute safety procedures and training. Further, it can be postulated that the client is receptive, but the failure has been one by the Canadian engineer to fulfill his contractual obligation; if so, professional ethics is a consideration.

Summary

This suggested approach is one way to analyze a case but is not the only way. What is important is to be systematic.

By some of the arguments advanced in this analysis, it is reasonable to conclude that, in the situation reported, there is no convincing evidence of any violation of the Code of Ethics even though the working conditions are definitely undesirable and unacceptable by Canadian standards. But when the situation is considered from another point of view, ethics is indeed a significant factor. It may not make much difference which argument prevails; in either case the same action would be appropriate and could be taken. That action would include persuasion, cooperation and training.

Safety of the local workers and of the Canadian engineer himself and the expectation that the completed project would be of a higher quality, and consequently a better reputation-builder, would be the incentive to stimulate the action.

ANSWERS TO SOME QUESTIONS

As stated previously, wherever appropriate in the context, the singular is intended to include the plural and the masculine to include the feminine.

Because of the nature of the subject and the broad scope of the considerations relevant to the situation presented, the answer given is not always the only acceptable answer to the question.

1. (a) No, the society has its own rules and is not bound by a code of ethics of some other association.

   However, it would have been appropriate and surely it would have been the polite thing to do to give some credit to the senior author, even though he was not eligible for an award because he was not a member of that particular engineering society at that time.

   (b) The same comments apply to the society officers as to the society in (a) above. Perhaps an argument can be advanced that there was no obligation because of the rule of the society, but certainly it would have been logical and courteous to give some credit to the senior author for his part in the research. This reasoning considers that the "senior" author was not just "older" than the other two but that he did have a major role in the project on which the research paper was based.

   Continuing the assumption that he had been an active member of the research team, certainly the two award recipients did have an obligation to give credit to their colleague. See B.C. Code, Section 14.

2. Before being able to determine whether B acted unethically by undertaking to do the work, it is necessary to know whether A's connection with the work had been terminated. Because the statements by the owner and by engineer A about termination of agreement were contradictory, B had an obligation to find out which statement was true before he undertook the work. What were the facts?

   Loyalty to a confrère might sway B's judgment towards believing engineer A's story, while wishful thinking might make the owner's version more attractive and so easy for B to believe, especially if he wanted to work badly enough. However, neither loyalty nor wishful thinking is a sound substitute for critical examination. Under circumstances such as those cited, it may be necessary to retain proper legal counsel to interpret the agreement in question. If there has been valid termination of the agreement between the owner and A, it was ethical for B to undertake the work; if the termination was not valid then it was unethical for B to do what he did. See APEGGA Code 15.

3. There are two points to be considered in this question and we should deal with them separately.

   One point is whether the standard is proper; the other is whether the products meet the requirements of that standard.

   John Doe's assumption that his signature on a report that stated that the samples conformed with the requirements of the standard would indicate that he endorsed the standard is incorrect. It would indicate that the samples conformed with the requirements of the standard, and that only.

   What he should do is get on with the job, conduct the tests and report on them.

   Further, if he feels strongly enough that the standard should be reviewed, and perhaps revised, he should take the appropriate steps that are provided for that to be done.

4. To determine the answer to this question we must consider what different factors are involved and the various possibilities.

   When the engineer-consultant was working with the government there was a requirement for confidentiality of information between consultant and client. This is no longer a factor since the report has been published and is available to the public.

   One might take the position that not being aware that the report is available is evidence that the company is in need of, and should retain, a consultant to bring it up to date on such matters. If this is so, why bother with such details as a two-year-old report? There was a requirement for confidentiality of information between consultant and client until the project was terminated. This is no longer a factor since the report has been published and is available to the public.

Another possible situation is that, during the time between that when the study was made for the government and now, there may have been substantial advances in the specialty. One must realize that "the company" per se is difficult to deal with; any deal must be made with some person who is acting for the company. It is not improbable that the officer of the company who had communicated with this engineer was a top-flight metallurgical engineer who was familiar with the study the consultant had made for the government. The impression created would be not too favourable if the consultant appeared in any way to be withholding information by not mentioning the earlier report.

Even without the careful consideration of these various relevant factors, the answer seemed obvious at the start. Yes, he should inform the company of the availability of the report before he accepts the assignment; honesty and integrity must predominate. See Newfoundland Code, Item 3.(f).

5. The question states that the competing similar product is of lower quality at lower cost, and that the engineers at XL Co. were instructed to redesign their product so that it may be made available to the market at a lower price.

Much depends on how we interpret the words "made available at a lower price." We may speculate that the redesigning of the XL product would introduce a safety hazard but there is nothing in the wording of the question to suggest that that would be so.

It may be that the original design was unduly elaborate and perhaps the most significant effect of redesigning the product would be to simplify the manufacturing process and result in a product that not only would cost less to produce (and consequently could be made available at a lower price) but one that was a better product.

Because there is no convincing evidence that there is any sound reason the instructions "to redesign the product" should not be carried out, we conclude that the engineers *do not* have "a proper interest and ethical obligation to protest the company's decision" nor to refuse to follow the instructions they were given to redesign the product so that is could be made available at a lower price.

9. This question describes a situation where the architect has contracted with the owner to provide certain services. Also, the structural engineer has contracted with the architect to provide certain of those services. While not stated explicitly in the question, the usual arrangement is that there is no contractual relationship between the owner and the engineer and I have assumed that to be so in this case.

   The matters to be considered in this situation are more than the ethics; they include strategy in interprofessional relations, tactics in negotiating, marketing of consulting engineering services and the law of contract.

   One significant question is: "In the contract between the architect and the structural engineer, is the signing of the certificate of compliance listed as one of the services to be provided?"

   If so, it would be unethical for the engineer not to provide the certificate because his work would not have been completed until he did.

   If, however, signing that certificate had not been listed but had been considered to be an incidental detail, it would *not* be unethical for him to refuse to execute the certificate because he had not been paid for his services. [Code 5(f).] Whether that refusal would be good business tactics is another matter, which he should consider. In this respect one must realize that any delay in the collection of fees is a much more important factor in 1988 than it was in 1918 because interest rates now are much higher than they were then.

10. We must consider different features of the proposal; some have ethical considerations while others are more a matter of business practice. Because the expression "value engineering" may mean different things to different people, to answer this question we must analyze the proposal carefully.

    The proposal is that engineering services will be provided on a "contingent-fee" basis. That is: if the client is not satisfied with and does not adopt the recommended revisions in design and/or procedures, no fee will be charged.

One could argue that this contravenes the codes of ethics, which state that a professional engineer shall uphold the principle of adequate compensation for engineering work. The rebuttal to that argument is that the prospect of the "portion of the saving" to be paid to the engineer by the client is "adequate compensation," and, further, that the engineer is prepared to bank on the worth of the "revisions in the design or production procedures" that would be suggested by his report.

Another comment might be made: that while the proposed arrangement may not be unethical, it is not sound business practice. That point of view could be disputed. This somewhat unusual approach might be the start of a successful and continuing client-engineer relationship.

In conclusion, the proposal is *not* unethical.

11. The status and authority of engineer B with his employer, the industrial corporation, is not stated explicitly in the question and I have interpreted the situation to be that the officers of the corporation are aware of and approve of him being chairman of that civic committee and spending time on such an activity. It is quite probable that such participation in civic affairs is of benefit to the public image of the corporation.

    Under these conditions:

    (a) The instructions given by engineer B to engineer A would not violate the APEO Code of Ethics, Section 91.1 nor Misconduct 86.(b)(c)(d) and (g).

    (b) It would *not* be unethical for engineer A to carry out these instructions. He should, however, take whatever steps are necessary to ensure that the architect is informed, either by engineer B or by the secretary of the civic committee, that his plans and specifications are to be reviewed for the purposes stated. A prudent engineer A would require that this information be transmitted in writing and that he receive a copy of the written communication, which he should keep in his files. Also, he should take whatever action may be necessary to be informed, by the proper authority or officer within the corporation, that it will be in order for him to spend his time (time paid for by the corporation) on this assignment. Code 1(a) and (c).

14. The question states that public safety would be endangered by a failure of the system, if one were to oc-

cur. The work for which the engineer was responsible was such that he had an obligation to make effective provision for the safety of the life and health of any person who might be affected by that work; and to act to correct or report any situation which he felt might endanger the public. See Saskatchewan Code, 2.(e).

Because the assembly failed to pass the final tests, the engineer had an obligation to report this to his superior, and it is stated in the question that he did that.

When told that the equipment had been shipped without the purchaser being informed of the unsatisfactory test result, the engineer is faced with an ethical problem.

There is some probability that the assembly test was not appropriate and that the system was indeed suitable and safe, but it is more probable that such a speculation is just wishful thinking and that the poor performance of the assembly in the test was because of some deficiency in the assembly.

Shipment having already been made, the engineer demands of his management that the purchaser be informed, at least, that satisfactory shop-test results have been obtained only for the components, and that the purchaser's acceptance test of the system must be made before public safety can be considered covered. The purchaser must be informed that the shop test of the assembly was unsatisfactory, saying in what respect.

The engineer is bound to insist that his company do no less than this; the company is going to be liable for any additions or alterations to the system anyway, if these have to be made to meet the contract specifications. To thus make a virtue of necessity is good business for the company, and the action is vital to the reputation of the engineer.

If the company refuses to so warn the purchaser, the engineer will have to tell his employer that he is ethically bound to report the matter. At this stage he should report to his superior in writing, with a further statement that, if proper action is not taken by his employer, he will report the situation to an occupational health and safety inspector of the ministry of labour, responsible for the safety of such systems. The employer's refusal is unlikely. If the employer persists in refusal, the engineer will have to do what he says he will do. If his employment is terminated, he could probably bring action for wrongful dismissal.

18. The main considerations in the context of the situation reported are the meanings of the words "development," "ethics" and "knowledge."

    Included in the definition of an engineer we find "one who contrives, designs," and so the development of the processing equipment is engineering work. Sections 91.3 of the Code of Ethics and 86.(2)(f) Misconduct of O.Reg. 538/84 may be applicable as the engineer progresses with the design of the equipment.

    Any knowledge acquired by a person is his and he may make use of that knowledge if he wishes to, provided that by so doing he does not contravene some code, custom, law or regulation.

    It is usual for an engineer to be employed because of experience he has had. It is hoped and expected that the knowledge and training he obtained through that experience will have made him more capable, and hence more valuable to his employer.

    Because this is a usual practice we conclude that it is "accepted practice" and that, yes, the engineer may ethically apply *his* knowledge to the development of equipment for his present employer. Sections 91.3 and 86(2)(f) referred to above include some points that the engineer should bring to the attention of Company A, his present employer.

    The question states that "the technical information concerning the equipment has not been published in the technical press, or otherwise released," but we do not know whether Company B has patented or otherwise imposed any effective legal restrictions on the use of any of the features of the processing equipment.

    We must realize that while the engineer may ethically apply *his knowledge* to the development of equipment for his present employer, he has an obligation to bring to the attention of his present employer the possibility of the existence of such restrictions and what could be the consequences of violating them if they do exist.

21. The situation reported is not unusual in consulting practices. The three parts to the questions are whether each of three different features of the situation is ethical or

unethical. Whether a feature is usual or unusual may be of interest to us but has little, if any, relevance to the questions asked.

(a) We shall defer answering this part until after parts (b) and (c) have been answered, because the answers to them provide useful input in considering this one.

(b) In the profession/business of consulting engineering, many different services are performed and several of them do not require engineers. Among these are: some scientific studies; public-relations activities; financial controls; legal matters; secretarial duties; staff recruitment. It is logical to consider the locating and arranging assignments for the firm to be another such service that someone who is not a professional engineer could do quite properly. That is not to say that the person might not be a professional engineer, but that he need not be. Considering the facts stated, it is evident that this is the business arrangement between the firm and the representative and no evidence has been presented to suggest that both parties are not in agreement with that arrangement.

The codes of ethics require that "a P.Eng. shall uphold the principle of adequate compensation for engineering work." It should be noted, however, that this employee is not a professional engineer and so is not subject to the rules of the Code, and further that there is no evidence that he does not consider the compensation to be adequate. After all, he agreed to the business deal; otherwise he wouldn't be working at the job.

And so the answer to this part of the question is: No, it is not unethical to employ a non-engineer in this role, but it should be noted that the non-engineer should not be misrepresented as being an engineer.

(c) Our codes of ethics stipulate that "a P.Eng. shall not attempt to gain an advantage . . . by paying or accepting a commission." Because "he is paid a salary plus an achievement bonus," the arrangement would then be viewed in an entirely different light.

This is a fine example of the ambiguity of certain words in the English language. It is much more reasonable to consider the compensation package to be "a salary plus an achievement bonus." And so: No, it is not unethical to compensate such a representative on an achievement-bonus basis (sometimes called a commission).

(a) Considering the analyses and answers in (b) and (c), it follows that the answer to (a) is: Yes, it is ethical for an engineering firm to have a non-engineer as a representative to discuss engineering aspects of a project and contract negotiations with a client, provided that he is not misrepresented as being an engineer.

23. No, it was not acceptable professional practice. There was a real danger that changes would be made to what was shown on the tracings and that some of these changes would result in some features of the framework and scaffolding not being as required by the safety legislation. Because the engineer no longer had control of the drawings, the contractor could have made changes for main material and connection details and the engineer could be held responsible for these changes because his seal was on them. The engineer might have been able, by some cumbersome procedures, to establish that the changes had been made without his authority or permission and consequently that he should not be held responsible for them, but perhaps he could not escape the responsibility. Certainly it was not sound practice.

After he had inspected the scaffolding and had learned that in many significant parts his design had been ignored, he should investigate those parts that did not conform with his design. Just because a detail differed from his design would not be sufficient reason to reject it but the engineer would have to be satisfied that, even if different, that way of doing it met the safety requirements so as to ensure the safety of the workers. The engineer should inform his client, the contractor, in writing, that the placing of the concrete should not be started before all these changes in the formwork and scaffolding, which he had listed in his review, had been completed, reinspected and accepted by him.

24. It is stated in the question that the three main functions of the company are consulting-engineering services, construction, and manufacturing; and that all of the engineering operations are directed by registered professional engineers. Undoubtedly, such an arrangement could be

abused and there would be opportunities for unprofessional action and collusion because of the common ownership. That is not to say, however, that there would be such behaviour.

In fact, such opportunities for misbehaviour exist regardless of the ownership of an organization carrying on any one of those three functions. The practice of consulting engineering, as such, by this company is not a violation of the Code, but there is always the possibility that an engineer employed by this or any other company might violate the Code.

It should be noted that a company does not have any obligation under the Code of Ethics. The Code imposes obligations on a member of an association of professional engineers and he, in turn, may have some influence on the actions of his employer—the company.

26. (a) It was not ethical for the chief engineer to recommend the award of the contract that he did. See Québec Code 3.02.08. There must have been some bias in judgment or some undisclosed non-technical reason for the chief engineer to feel that the concrete design must be used rather than asphalt, notwithstanding "the understanding the two would be of equal quality and would give equal service." The instruction "to revise the concrete design to a lower standard but not to change the asphalt design," could reasonably be interpreted as evidence of a conflict of interest or of incompetence.

    (b) It was misconduct for the chief engineer to give the instructions he did to the engineers, to revise the concrete design only. See Québec Code, 3.02.10.

28. Of particular significance in this question is the statement that the engineer *knows* that a key ingredient in one of the products *has been linked to a large number of cancer cases.*

    We must consider how he became aware of that knowledge and whether the source of the information is dependable. While this is not what is asked, an obvious related question is: Why did not this young engineer do something about the situation before the company was asked by the inspector from the ministry of labour to provide the information stipulated? Our codes of ethics stipulate that a person in a position of trust, such as this young engineer is, should be mindful of public health requirements.

The proper and necessary action for this engineer to take is to prepare his report stating the "facts" as he understands them to be and submit the report to the person who ordered him to prepare it. The report should provide the information requested and should include a statement that one of the ingredients in one of the products has been linked to a large number of cancer cases. Perhaps the linkage was mere speculation by some undependable publicity-seeking investigator, but perhaps it was a conclusion after a thorough amd meticulous study by a competent investigator; the wording of the question does not clarify this point.

The engineer should anticipate a situation where that person objects to the reference to the "fact" or "suspicion" that the ingredient in question has been linked to a large number of cancer cases and may wish to have that omitted from the report. If that situation arises, the engineer must insist that this reference be retained in his report. If this is not done he must report the information directly to the appropriate officer in the ministry of labour, even though such action by him may mean that he will be dismissed by his employer.

29. (a) Engineering work and those who do that work make up a large part of today's industrial activity. Because of the importance of this, anything that will enhance the quality of that work and the dependability of those who do it is of definite value to the industry. A professional engineering association expects its members to respect the rules of its Code of Ethics and administers the Act in such a way as to achieve this. As a result, most members of the association do behave in accordance with the spirit of the Code; this has brought about a situation where the Code is considered to be a "standard" for ethical behaviour by professional engineers. For this reason, aside from industry's dependence on the professional engineer, it owes much to his Code of Ethics for what it has done to enhance the dependability of the P.Eng. who is such an important factor in the industrial process.

(b) There are two aspects to be considered in response to this question.

One of them has legal implications: If, indeed, the meaning of "engineering positions" is that the law at the place of work requires that these positions be held by registered professional engineers, the situation is illegal and subject to the applicable statutory penalties.

The other is concern with the ability of the persons to do the work. The technology of the work to be done may be such that for the work to be done properly anyone doing it must have the qualifications that are required for admission to an association of professional engineers. Technical or technological incompetence could have a serious, adverse effect on the company's operations, because without his competence there would be a greater probability of a lack of appreciation of such things as the significance of obsolescence in a manufacturing process. This is just one of many examples of the adverse effects of the lack of adequate engineering input.

(c) The brief answer to this is yes, in the case of reputable industrial firms, which means, fortunately, most firms in Canada, and the reasons are as presented more fully in response to (a). There are, of course, unusual situations where those conditions and relationships do not exist. At times a fly-by-night industry will have shady—perhaps even illegal—objectives that are in conflict with those of the Code of Ethics.

31. See Manitoba Code, Item 2, Duty to the State. If the engineer were to disclose to a court of law that the fee for his services (read "evidence") was to be contingent upon the decision of the court, it would quite definitely classify his evidence and his opinion as prejudiced. For this reason, the answer to the question is simply no, it would not be ethical for an engineer to serve as an expert witness in a lawsuit under the conditions described in the question, especially if he tried to keep secret his fee arrangement.

32. "Moonlighting" is the word used to describe the situation presented in this question. As in so many of the cases where there is moonlighting, it would be unusual if A and B were to agree that, yes, it was unethical, or, no, it was not unethical, for A to employ B's sub-professional employees under these circumstances.

The APEO manual of professional practice deals with moonlighting by professional engineer employees but not by non-professionals. In the situation referred to in this question, we are told that B is of the opinion that the moonlighting by his employees causes them to be less productive while working for him than they would be if they were not doing that after-hours work. Whether that lower productivity is real or just fancied by B would be difficult, if not impossible, to determine.

The question states that B *learned* that some of his employees were doing this work with A, so we may conclude that A did not tell him.

In view of this assessment of the situation, yes, it was unethical of A to employ B's sub-professional staff under the circumstances stated.

38. Facts:

You live in Brownsville, a small community;

Disneyworld has engaged your firm to undertake a study of the impact of an entertainment park near your home;

For several and varied personal reasons you would be displeased if the park were built there;

Your firm has asked you to be the project manager of this study;

Question: Can you undertake the assignment?

Answer:

Yes—even on a sunny day, some rain may fall. The only proviso is that you must be convinced that you can do a sound engineering study of the environmental impact of the park in the community and avoid having your personal desires and prejudices distort your analysis and report. It is reasonable to conclude that the officers of your firm selected you for the study, not because you lived in the community, but because they considered that you would do the job well.

40. Facts:

You are the Chief Engineer of company XYZ which does business in Ontario and you hired Mr. A for an engineering position on your staff;

Mr. A's statements and behaviour led you to believe that he was either a member of, or was licensed to practise by, APEO;

When you learned that Mr. A did not have a licence to practise in Ontario you terminated his employment with XYZ immediately.

Question 1: Was your action ethical?

Question 2: Was Mr. A's action unethical?

Answer:

Item 91.1.i of the APEO Code of Ethics states that: "It is the duty of a practitioner to the public, to his employer, to his clients, to other members of his profession, and to himself to act at all times with fairness and loyalty to his associates, employers, clients, subordinates and employees."

Item 91.8. states that: "A practitioner shall maintain the honour and integrity of his profession and without fear or favour expose before the proper tribunals unprofessional, dishonest or unethical conduct by any other practitioner."

The answer to Question 1 is yes, your action was ethical, based on item 91.8 of the Code of Ethics. However, it is relevant to consider code item 91.1.i and the possibility that Mr. A was not aware that his membership in the Ordre did not, per se, give him the rights and privileges of a professional engineer in Ontario.

The answer to Question 2 is yes, Mr. A's action was unethical: ignorance of the law is no excuse.

APPENDIX

The Sons of Martha

Codes of Ethics of All the Provincial and Territorial Associations of Professional Engineers in Canada

THE SONS OF MARTHA

The Sons of Mary seldom bother, for they have inherited
that good part;
But the Sons of Martha favour their Mother of the careful
soul and the troubled heart.
And because she lost her temper once, and because she was
rude to the Lord her Guest,
Her Sons must wait upon Mary's Sons, world without end,
reprieve, or rest.

It is their care in all the ages to take the buffet and cushion
the shock,
It is their care that the gear engages; it is their care that the
switches lock.
It is their care that the wheels run truly; it is their care to
embark and entrain,
Tally, transport, and deliver duly the Sons of Mary by land
and main.

They say to mountains, "Be ye removed." They say to the
lesser floods, "Be dry."
Under their rods are the rocks reproved—they are not
afraid of that which is high.
Then do the hill-tops shake to the summit—then is the bed
of the deep laid bare,
That the Sons of Mary may overcome it, pleasantly sleep-
ing and unaware.

They finger death at their gloves' end where they piece and
repiece the living wires.
He rears against the gates they tend: they feed him hungry
behind their fires.
Early at dawn, ere men see clear, they stumble into his
terrible stall,
And hale him forth like a haltered steer, and goad and turn
him till evenfall.

To these from birth is Belief forbidden, from these till
death is Relief afar.
They are concerned with matters hidden—under the earth-
line their altars are—
The secret fountains to follow up, waters withdrawn to re-
store to the mouth,
And gather the floods as in a cup, and pour them again at
a city's drouth.

They do not preach that their God will rouse them a little
before the nuts work loose.
They do not teach that His Pity allows them to drop their
job when they dam'-well choose.
As in the thronged and the lighted ways, so in the dark and
the desert they stand,
Wary and watchful all their days that their brethren's days
may be long in the land.

Raise ye the stone or cleave the wood to make a path more
fair or flat—
Lo, it is black already with blood some Son of Martha
spilled for that!
Not as a ladder from earth to Heaven, not as a witness to
any creed,
But simple service simply given to his own kind in their
common need.

And the Sons of Mary smile and are blessed—they know
the Angels are on their side.
They know in them is the Grace confessed, and for them
are the Mercies multiplied.
They sit at the Feet—they hear the Word—they see how
truly the Promise runs.
They have cast their burden upon the Lord, and—the Lord
He lays it on Martha's Sons!

Rudyard Kipling chose his poem "The Sons of Martha," written in 1907, to describe dramatically the great responsibilities of engineers.

CODES/CANONS OF ETHICS

There is much more to the ethical aspects of professional engineering practice than the rules listed in a code of ethics.

Just as it would be absurd to believe that you were a mathematician because you could recite the multiplication table, or even if you were competent to punch the proper keys on a computer in the proper sequence to get the answer to a particular problem; so also it would be absurd to believe that you had a thorough and proper comprehension of the ethical aspects of professional engineering practice because you had memorized and could recite a code of ethics.

Not only must a professional engineer know what the Code requires but he must be aware of, and comprehend, the significance of its various stipulations. Each of the professional engineering associations in Canada has developed a Code of Ethics. While there are some differences, there is general agreement among these codes.

This Appendix contains the code or canons of ethics of each of the eleven professional engineering associations and the Ordre in all of the provinces and territories in Canada, listed alphabetically from Alberta to the Yukon. The section of the Ontario Regulation entitled Professional Misconduct is also included.

In each case permission to publish the material has been granted by the association or the Ordre.

ASSOCIATION OF PROFESSIONAL ENGINEERS, GEOLOGISTS AND GEOPHYSICISTS OF ALBERTA

CODE OF ETHICS

1. A professional engineer, geologist or geophysicist shall recognize that professional ethics are founded upon integrity, competence and devotion to service and to the advancement of human welfare. This concept shall guide his conduct at all times.

DUTIES OF THE PROFESSIONAL ENGINEER, GEOLOGIST, OR GEOPHYSICIST TO THE PUBLIC

A professional engineer, geologist or geophysicist shall

2. have proper regard in all his work for the safety, health and welfare of the public;
3. endeavour to extend public understanding of engineering, geology and geophysics and their places in society;
4. where his professional knowledge may benefit the public, seek opportunities to serve in public affairs;
5. not be associated with enterprises contrary to the public interest or sponsored by persons of questionable integrity;
6. undertake only such work as he is competent to perform by virtue of his training and experience;
7. sign and seal only such plans, documents or work as he himself has prepared or carried out or as have been prepared or carried out under his direct professional supervision;
8. shall express opinions on engineering, geological or geophysical matters only on the basis of adequate knowledge and honest conviction.

DUTIES OF THE PROFESSIONAL ENGINEER, GEOLOGIST OR GEOPHYSICIST TO HIS CLIENT OR EMPLOYER

A professional engineer, geologist or geophysicist shall

9. act for his client or employer as a faithful agent or trustee;
10. not accept remuneration for services rendered other than from his client or his employer;

11. not disclose confidential information without the consent of his client or employer;
12. not undertake any assignment which may create a conflict of interest with his client or employer without the full knowledge of the client or employer;
13. present clearly to his clients or employers the consequences to be expected if his professional judgment is overruled by other authorities in matters pertaining to work for which he is professionally responsible.

DUTIES OF THE PROFESSIONAL ENGINEER, GEOLOGIST OR GEOPHYSICIST TO THE PROFESSION

A professional engineer, geologist or geophysicist shall

14. endeavour at all times to improve the competence, dignity and prestige of his profession;
15. conduct himself towards other professional engineers, geologists and geophysicists with fairness and good faith;
16. not advertise his professional services in self-laudatory language or in any other manner derogatory to the dignity of his profession;
17. not attempt to supplant another engineer, geologist or geophysicist in an engagement after definite steps have been taken toward the other's employment;
18. not use advantages of a salaried position to compete unfairly with another engineer, geologist or geophysicist;
19. not exert undue influence or offer, solicit or accept compensation for the purpose of affecting negotiations for an engagement;
20. build his professional reputation on the basis of the merit of services offered and shall not compete unfairly with others or compete primarily on the basis of fee, without due consideration for other factors;
21. advise the Council of any practice by another member of his profession which he believes to be contrary to this Code of Ethics.

Latest Revision Date Shown 1978

ASSOCIATION OF PROFESSIONAL ENGINEERS OF BRITISH COLUMBIA

CODE OF ETHICS

14. Preamble—The following is prescribed as the Code of Ethics of the Association, and the Engineer is bound by its provisions just as he is bound by the provisions of the Engineers Act, 1979 and by the Bylaws of the Association.

The professional engineer shall act at all times with fairness, loyalty and courtesy to his associates, employers, employees and clients, and with fidelity to the public needs. He shall approach his work with devotion to high ideals, personal honour and integrity.

The purpose of the Code is two-fold:

(1) To give general statements of the principles of honourable conduct which, over the years, members of the profession of engineering have come to accept as required of each member in order that he may fulfill his duty to the public, to the profession, and to his fellow members.

(2) To give some specifics in the sub-sections, both of required standards and prohibited actions, in order that they may act as a guide, to the intent of the general statements. These specifics, it is emphasized, are only some examples of the broad principles upon which members of this profession must appraise and govern their own conduct.

The following Code of Ethics is promulgated as a general guide and not as a denial of the existence of other duties equally imperative, but not specifically included.

SECTION 1

The Engineer will be guided in all his professional relations by the highest standards of integrity.

(a) He will be realistic and honest in the preparation of all estimates, reports, statements and testimony.

(b) He will not distort or alter facts in an attempt to justify his decisions or avoid his responsibilities.

(c) He will advise his client or employer when he believes a project will not be successful or in the best interests of his client or his employer or the public.

(d) He will not engage in any work outside his salaried work to an extent prejudicial to his salaried position.

(e) In the interpretation of contract documents, he will maintain an attitude of scrupulous impartiality as between parties and will, as far as he can, ensure that each party to the contract will discharge the duties and enjoy the rights set down in the contract agreement.
(f) He will not use his professional position to secure special concessions or benefits which are detrimental to the public, his clients or his employer.

SECTION 2

The engineer will have proper regard for the safety, health and welfare of the public in the performance of his professional duties. He will regard his duty to the public safety and health as paramount.
(a) He will guard against conditions that are dangerous or threatening to life, limb or property on work for which he is responsible, or if he is not responsible will properly call such conditions to the attention of those who are responsible.
(b) He will present clearly the consequences to be expected if his engineering judgment is overruled.
(c) He will seek opportunities to work for the advancement of the safety, health and welfare of his community.
(d) He will guard against conditions which are dangerous or threatening to the environment and he will seek to ensure that all standards required by law for environmental control are met.

SECTION 3

The Engineer may promote and advertise his work or abilities provided that:
(a) The advertising preserves the public interest by reporting accurate and factual information which neither exaggerates nor misleads.
(b) The advertising does not impair the dignity of the profession.
(c) Statements do not convey criticism of other engineers directly or indirectly.

SECTION 4

The Engineer will endeavor to extend public knowledge and appreciation of engineering and its achievements and will endeavor to protect the engineering profession from misrepresentation and misunderstanding.

(a) He will not issue statements, criticisms, or arguments on engineering matters connected with public policy which are inspired or paid for by private interests, unless he indicates on whose behalf he is making the statement.

SECTION 5

The Engineer may express an opinion on an engineering subject only when founded on adequate knowledge and honest conviction.

(a) In reference to an engineering project in a group discussion or public forum, he will strive for the use of pertinent facts, but if it becomes apparent to the engineer that such facts are being distorted or ignored, he should publicly disassociate himself from the group or forum.

SECTION 6

The Engineer will undertake engineering assignments for which he will be responsible only when qualified by training or experience: and he will engage, or advise engaging, experts and specialists whenever the client's or employer's interests are best served by such service.

(a) He will not sign or seal plans, specifications, reports or parts thereof unless actually prepared by him or prepared under his supervision.

SECTION 7

The Engineer will not disclose confidential information concerning the business affairs or technical processes of any present or former client or employer without his consent.

SECTION 8

The Engineer will endeavor to avoid a conflict of interest with his employer or client, but when such conflict is unavoidable, the Engineer will fully disclose the circumstances to his employer or client.

(a) He will inform his client or employer of any business connections, interests, or circumstances which may be deemed as influencing his judgment or the quality of his services to his client or employer.

(b) He, while a member of any public body, will not act as a vendor of goods or services to that body.

SECTION 9

The Engineer will uphold the principle of appropriate and adequate compensation for those engaged in engineering work.

(a) He will not normally undertake or agree to perform any engineering service on a free basis, except for civic, charitable, religious, or nonprofit organizations when the professional services are advisory in nature.

(b) He will not compete improperly by reducing his usual charges to underbid a fellow engineer after having been informed of that engineer's charge.

SECTION 10

The Engineer will not accept compensation, financial or otherwise, from more than one interested party for the same service, or for services pertaining to the same work, unless there is full disclosure to and consent of all interested parties.

(a) He will not accept financial or other considerations, including free engineering designs, from material or equipment suppliers as a reward for specifying their product.

(b) He will not accept commissions or allowances, directly or indirectly, from contractors or other parties dealing with his clients or employer in connection with work for which he is responsible.

SECTION 11

The Engineer will not compete unfairly with another engineer by attempting to obtain employment or advancement or professional engagements by taking advantage of a salaried position, or by criticizing other engineers or by other improper or questionable methods.

(a) He will not attempt to supplant another engineer in a particular employment after becoming aware that definite steps have been taken toward the other's employment.

(b) He will not offer to pay, or agree to pay either directly or indirectly, any commission, political contribution, gift, or other consideration in order to secure work.

(c) He will not solicit or accept an engineering engagement on a contingent fee basis if payment depends on a finding of economic feasibility or other preconceived conclusion.

SECTION 12

The Engineer will not attempt to injure maliciously or falsely, directly or indirectly, the professional reputation, prospects or practice of another person.

(a) He will not accept any engagement to review the work of a fellow engineer except with the knowledge of and after communication with such fellow engineer, where such communication is possible.

(b) He will refrain from expressing publicly an opinion on an engineering subject unless he is informed as to the facts relating thereto.

(c) Unless he is convinced that his responsibility to the community requires him to do so, he will not express professional opinions which reflect on the ability or integrity of another person or organization.

(d) He will exercise due restraint in his comments on another engineer's work.

(e) If he considers that an engineer is guilty of unethical, illegal or unfair practice, he will present the information to the Registrar of the Association.

(f) An engineer is entitled to make engineering comparisons of the products offered by various suppliers.

SECTION 13

The Engineer will not associate with or allow the use of his name by an enterprise of questionable character, or by one which is known to engage in unethical practice.

(a) He will not use association with a non-engineer, a corporation, or partnership as a "cloak" for unethical acts, but must accept personal responsibility for his professional acts.

SECTION 14

The Engineer will give credit for engineering work to those to whom credit is due, and will recognize the proprietary interests of others.

(a) Whenever possible, he will name the person or persons who may be individually responsible for designs, inventions, writings, or other accomplishments.

(b) When an engineer uses designs supplied to him by a client or by a consultant, the designs remain the property of the client or consultant and should not be duplicated by the engineer for others without express permission.

(c) Before undertaking work for others in connection with which he may make improvements, plans, designs, inventions, or other records which may justify copyrights or patents, the engineer should enter into a positive agreement regarding the ownership of such copyrights and patents.

SECTION 15

The Engineer will co-operate in extending the effectiveness of the profession by interchanging information and experience with other engineers and students, and will endeavor to provide opportunity for the professional development and advancement of engineers in his employ or under his supervision.

(a) He will encourage his engineering employees in their efforts to improve their education.

(b) He will encourage engineering employees to attend and present papers at professional and technical society meetings.

(c) He will urge his qualified engineering employees to become registered.

(d) He will assign a professional engineer duties of a nature to utilize his full training and experience, insofar as possible.

(e) He will endeavor to provide a prospective engineering employee with complete information on working conditions and his proposed status of employment, and after employment will keep him informed of any changes in them.

SECTION 16

The Engineer will observe the rules of professional conduct which apply in the country in which he may practise. If there be no such rules, then he will observe those set out by this code.

INTERPRETATION

15. In the event of any dispute as to the meaning or intent of these Bylaws, the interpretation of the Council shall be final, subject to the right of Appeal as contained in Section 32 of the Act.

Where the word "Act" appears in the foregoing Bylaws, it shall include the Engineers Act and all subsequent Amending Acts, unless the context otherwise requires.

REPEAL OF OLD BYLAWS

16. Upon the coming into force of the foregoing Bylaws, all the Bylaws of the Association previously in force shall stand revoked.

Latest Revision Date Shown 1979.

ASSOCIATION OF PROFESSIONAL ENGINEERS OF THE PROVINCE OF MANITOBA

PROFESSIONAL ENGINEERS CODE OF ETHICS

1. PREAMBLE

1.1 Honesty, justice and courtesy form a moral philosophy which, associated with mutual interest among men, constitute the foundation of ethics. The professional engineer should recognize such a standard, not in passive observance, but as a set of dynamic principles guiding his conduct and way of life. It is his duty to practise his profession according to this code of ethics.

1.2 As the keystone of professional conduct is integrity, the professional engineer will discharge his duties with fidelity to the public, his employers, and clients, and with fairness and impartiality to all. It is his duty to interest himself in public welfare, and to be ready to apply his special knowledge for the benefit of mankind. He should uphold the honour and dignity of his profession and also avoid association with any enterprise of questionable character. In his dealings with fellow engineers he should be fair and tolerant.

2. DUTY TO THE STATE

2.1 The professional engineer owes a duty to the state, to maintain its integrity and its law.

2.2 He shall at all times act with candour and fairness when engaged as an expert witness and give, to the best of his knowledge and ability, an honest opinion based on adequate study of the matter in hand.

3. DUTY TO THE PUBLIC

3.1 The professional engineer shall regard the physical and economic well-being of the public as his first responsibility in all aspects of his work.

3.2 He shall advise on, design or supervise only such projects as his training, ability and experience render him professionally competent to undertake.

3.3 He shall guard against conditions that are dangerous or threatening to health, life, limb or property on work for which he is responsible, or if he is not responsible, shall promptly call such conditions to the attention of those who are responsible.

3.4 He shall not associate himself with, or allow the use of his name by, an enterprise of doubtful character.

3.5 He shall not issue one-sided statements, criticism, or arguments on engineering matters which are initiated or paid for by parties with special interests, unless he indicates on whose behalf he is making the statement.

3.6 He shall refrain from expressing publicly an opinion on an engineering subject unless he is aware of all the related facts.

3.7 He shall not permit the publication of his reports, or parts of them in a manner calculated to mislead.

3.8 He shall on all occasions seal plans and specifications which legally require sealing and for which he is professionally responsible and ethically and legally entitled to seal, whether he acts in the capacity of a consultant or as an employee.

3.9 He shall sign or seal only those specifications and plans for which he is professionally responsible and which have been prepared by him or under his personal direction.

4. DUTY TO HIS CLIENT OR EMPLOYER

4.1 The professional engineer shall employ every resource of skill and knowledge that he commands to perform and satisfy the engineering needs of his client or employer in a truly worthy manner.

4.2 He shall act in professional matters for each client or employer as a faithful agent or trustee.

4.3 He shall ensure that the extent of his responsibility is fully understood by each client or employer before accepting a commission.

4.4 He shall not disclose any information concerning the business affairs or technical processes of clients or employers without their consent.

4.5 He shall engage, or advise each client or employer to engage, and shall co-operate with, other experts and specialists whenever the client's or employer's interests are best served by such service.

4.6 He shall present clearly the consequences to be expected from deviations proposed, if his engineering judgment is overruled by other authority, in cases where he is responsible for the technical adequacy of engineering work.

4.7 He shall inform his clients of any business connections, interest or circumstances which may be deemed as influencing his judgment or the quality of his services to his clients, before accepting a commission.

4.8 He shall not allow an interest in any business to adversely affect his decision regarding engineering work for which he is employed, or which he may be called upon to perform.

4.9 He shall not receive, directly or indirectly, any royalty, gratuity, or commission on any patented or protected article or process used in work upon which he is retained by his client, unless and until receipt of such royalty, gratuity or commission has been authorized in writing by his client.

4.10 He shall not accept compensation, financial or otherwise, from more than one interested party for the same service, or for services pertaining to the same work, without the consent of all interested parties.

4.11 He shall not be financially interested in the bids as or of a contractor on work for which he is employed as an engineer unless he has the written consent of his clients or employers.

4.12 He shall not accept commission, allowances, or fees, directly or indirectly, from contractors or other parties dealing with his client or employer in connection with work for which he is responsible, except proper engineering fees paid to him by the contractor for engineering work done prior to the award of the contract for a project for which the client subsequently appoints him consulting engineer, and then only if the client has been fully informed of the transaction prior to his appointment.

5. DUTY TO THE PROFESSION

5.1 The professional engineer shall think highly of his profession and its members, its history and traditions and shall act in a manner worthy of its honour and dignity at all times.

5.2 He shall constantly strive to broaden his knowledge and experience by keeping abreast of new techniques and developments in his field of endeavour and to maintain his reputation for skill and integrity.

5.3 He shall participate in extending the effectiveness of the engineering profession by interchanging information and experience with other engineers and students, and by contributing to the work of the engineering societies, schools and scientific and engineering press.

5.4 He shall not advertise in an unprofessional manner by making misleading statements regarding his qualifications or experience.

5.5 He shall endeavour to extend public knowledge of engineering, shall discourage the spreading of unfair or exaggerated statements regarding engineering, and shall strive to protect the engineering profession collectively and individually from misrepresentation and misunderstanding.

5.6 He shall present appropriate information to the registrar of the association, if he considers that a professional colleague is engaging in unethical, illegal or unfair practice.

6. DUTY TO HIS COLLEAGUES

6.1 The professional engineer shall take care that credit for engineering work is given to those to whom credit is properly due.

6.2 He shall uphold the principle of appropriate compensation for those engaged in engineering work, including those in subordinate capacities, as being in the public interest by maintaining the standards of the profession.

6.3 He shall endeavour to provide opportunity for the development and advancement of engineers and technical people in his employ.

6.4 He shall not attempt to injure falsely or maliciously, directly or indirectly, the professional reputation, prospects or business of another engineer.

6.5 He shall not accept any commission to review the work of a fellow engineer except with the knowledge of, and after communication with, such fellow engineer, where such communication is possible.

6.6 He shall refrain from criticizing another engineer's work in public, recognizing the fact that the engineering societies and the engineering press provide the proper forum for technical discussions and criticisms.

6.7 He shall not attempt to supplant another engineer after definite steps have been taken towards the other's employment.

6.8 He shall not compete with another engineer by reducing his usual fees or salary after having been informed of the other's fees or salary.

6.9 He shall not accept employment by a client, knowing that a claim for compensation or damages, or both, of a fellow engineer previously employed by the same client and whose employment has been terminated, remains unsatisfied, or until such claim has been referred to arbitration, or issue has been joined at law, or unless the engineer previously employed has neglected to press his claim legally, or the council of the association gives its assent.

6.10 He shall not use the advantages of a salaried position to compete unfairly with another engineer.

6.11 He shall always uphold the principle of a proper and separate rate fee being charged for engineering services for which he is responsible, whether acting as an independent consultant or as an employee of a firm supplying services in addition to those usually supplied by a consultant.

7. DUTY TO ALL CONTRACTUAL PARTIES

7.1 The professional engineer shall act with fairness and honesty with his client or employer and any party who engages in a contractual agreement with them.

7.2 He shall co-operate with any parties under contract and make every reasonable effort to facilitate completion of the work in accordance with the terms of the contract.

7.3 He shall act in an impartial manner in dealing with disputes between the contractual parties.

I hereby subscribe to the above code of ethics
to which I set my seal and signature

....P.Eng.

Adopted November 1st, 1921.
Rewritten and Adopted, February 28, 1968.
Latest Revision Date Shown 1968.
Reprinted by permission of The Association of
Professional Engineers of the Province of Manitoba.

ASSOCIATION OF PROFESSIONAL ENGINEERS OF THE PROVINCE OF NEW BRUNSWICK

CANONS OF ETHICS FOR ENGINEERS

FOREWORD

Honesty, justice, and courtesy form a moral philosophy which, associated with mutual interest among men, constitute the foundation of ethics. The engineer should recognize such a standard, not in passive observance, but as a set of dynamic principles guiding his conduct and way of life. It is his duty to practice his profession according to these Canons of Ethics.

As the keystone of professional conduct is integrity, the engineer will discharge his duties with fidelity to the public, his employers, and clients, and with fairness and impartiality to all. It is his duty to interest himself in public welfare, and to be ready to apply his special knowledge for the benefit of mankind. He should uphold the honor and dignity of his profession and also avoid association with any enterprise of questionable character. In his dealings with fellow engineers he should be fair and tolerant.

PROFESSIONAL LIFE

Sec. 1. The engineer will co-operate in extending the effectiveness of the engineering profession by interchanging information and experience with other engineers and students and by contributing to the work of engineering societies, schools, and the scientific and engineering press.

Sec. 2. He will not advertise his work or merit in a self-laudatory manner, and he will avoid all conduct or practice likely to discredit or do injury to the dignity and honor of his profession.

RELATIONS WITH THE PUBLIC

Sec. 3. The engineer will endeavor to extend public knowledge of engineering, and will discourage the spreading of untrue, unfair, and exaggerated statements regarding engineering.

Sec. 4. He will have due regard for the safety of life and health of the public and employees who may be affected by the work for which he is responsible.

Sec. 5. He will express an opinion only when it is founded on adequate knowledge and honest conviction while he is serving as a witness before a court, commission, or other tribunal.

Sec. 6. He will not issue ex parte statements, criticisms, or arguments on matters connected with public policy which are inspired or paid for by private interests, unless he indicates on whose behalf he is making the statement.

Sec. 7. He will refrain from expressing publicly an opinion on an engineering subject unless he is informed as to the facts relating thereto.

RELATIONS WITH CLIENTS AND EMPLOYERS

Sec. 8. The Engineer will act in professional matters for each client or employer as a faithful agent or trustee.

Sec. 9. He will act with fairness and justice between his client or employer and the contractor when dealing with contracts.

Sec. 10. He will make his status clear to his client or employer before undertaking an engagement if he may be called upon to decide on the use of inventions, apparatus, or any other thing in which he may have a financial interest.

Sec. 11. He will guard against conditions that are dangerous or threatening to life, limb, or property on work for which he is responsible, or if he is not responsible, will promptly call such conditions to the attention of those who are responsible.

Sec. 12. He will present clearly the consequences to be expected from deviations proposed if his engineering judgment is overruled by non-technical authority in cases where he is responsible for the technical adequacy of engineering work.

Sec. 13. He will engage, or advise his client or employer to engage, and he will co-operate with, other experts and specialists whenever the client's or employer's interests are best served by such service.

Sec. 14. He will disclose no information concerning the business affairs or technical processes of clients or employers without their consent.

Sec. 15. He will not accept compensation, financial or otherwise, from more than one interested party for the same service, or for services pertaining to the same work, without the consent of all interested parties.

Sec. 16. He will not accept commissions or allowances, directly or indirectly, from contractors or other parties dealing with his client or employer in connection with work for which he is responsible.

Sec. 17. He will not be financially interested in the bids as or of a contractor on competitive work for which he is employed as an engineer unless he has the consent of his client or employer.

Sec. 18. He will promptly disclose to his client or employer any interest in a business which may compete with or affect the business of his client or employer. He will not allow an interest in any business to affect his decision regarding engineering work for which he is employed, or which he may be called upon to perform.

RELATIONS WITH ENGINEERS

Sec. 19. The engineer will endeavor to protect the engineering profession collectively and individually from misrepresentation and misunderstanding.

Sec. 20. He will take care that credit for engineering work is given to those to whom credit is properly due.

Sec. 21. He will uphold the principle of appropriate and adequate compensation for those engaged in engineering work, including those in subordinate capacities, as being in the public interest and maintaining the standards of the profession.

Sec. 22. He will endeavor to provide opportunity for the professional development and advancement of engineers in his employ.

Sec. 23. He will not directly or indirectly injure the professional reputation, prospects, or practice of another engineer. However, if he considers that an engineer is guilty of unethical, illegal, or unfair practice, he will present the information to the proper authority for action.

Sec. 24. He will exercise due restraint in criticizing another engineer's work in public, recognizing the fact that the engineering societies and the engineering press provide the proper forum for technical discussions and criticism.

Sec. 25. He will not try to supplant another engineer in a particular employment after becoming aware that definite steps have been taken toward the other's employment.

Sec. 26. He will not compete with another engineer on the basis of charges for work by underbidding, through reducing his normal fees after having been informed of the charges named by the other.

Sec. 27. He will not use the advantages of a salaried position to compete unfairly with another engineer.

Sec. 28. He will not become associated in responsibility for work with engineers who do not conform to ethical practices.

Latest Revision Date Shown 1982.

ASSOCIATION OF PROFESSIONAL ENGINEERS OF THE PROVINCE OF NEWFOUNDLAND

CODES OF PROFESSIONAL ETHICS

GENERAL

1. A professional engineer owes certain duties to the public, his employers, to other members of his profession and to himself and shall act at all times with:
   (a) fairness and loyalty to his associates, employers, subordinates and employees;
   (b) fidelity to public needs; and
   (c) devotion to high ideals of personal honour and professional integrity.

DUTY OF PROFESSIONAL ENGINEER TO THE PUBLIC

2. A professional engineer shall,
   (a) endeavour at all times to enhance the public regard for his profession by extending the public knowledge thereof and discouraging untrue, unfair or exaggerated statements with respect to professional engineering;
   (b) not give opinions or make statements on professional engineering projects of public interest that are inspired or paid for by private interests unless he clearly discloses on whose behalf he is giving the opinions or making the statements;
   (c) not express publicly or while he is serving as a witness before a court, commission or other tribunal opinions on professional engineering matters that are not founded on adequate knowledge and honest conviction;
   (d) make effective provisions for the safety of life and health of a person who may be affected by the work for which he is responsible; and,
   (e) sign or seal only those plans, specifications and reports actually made by him or under his personal supervision and direction.

DUTY OF PROFESSIONAL ENGINEER TO EMPLOYER

3. A professional engineer shall,
   (a) act in professional engineering matters for each employer as a faithful agent or trustee and shall regard as confidential any information obtained by him as to the business affairs, technical methods or processes of an employer;
   (b) present clearly to his employers the consequences to be expected from any deviation proposed in the work if his professional engineering judgment is overruled by non-technical authority in cases where he is responsible for the technical adequacy of professional engineering work;
   (c) have no interest, direct or indirect, in any materials, supplies or equipment used by his employer or in any person or firms receiving contracts from his employer unless he informs his employer in advance of the nature of the interest;
   (d) not tender on competitive work upon which he may be acting as a professional engineer unless he first advises his employer;
   (e) not act as consulting engineer in respect of any work upon which he may be the contractor unless he first advises his employer; and,
   (f) not accept compensation, financial or otherwise, for a particular service, from more than one person except with the full knowledge of all interested parties.

DUTY OF PROFESSIONAL ENGINEER TO OTHER PROFESSIONAL ENGINEERS

4. A professional engineer shall,
   (a) conduct himself towards other professional engineers with courtesy and good faith;
   (b) not accept any engagement to review the work of another professional engineer for the same employer except with the knowledge of that engineer, or except where the connection of that engineer with the work has been terminated;
   (c) not maliciously injure the reputation or business of another professional engineer;
   (d) not attempt to gain an advantage over other members of his profession by paying or accepting a commission in securing professional engineering work; and,

(e) not advertise in a misleading manner or in a manner injurious to the dignity of his profession, but shall seek to advertise by establishing a well-merited reputation for personal capacity.

DUTY OF PROFESSIONAL ENGINEER TO HIMSELF

---

5. A professional engineer shall,
   (a) maintain the honour and integrity of his profession and without fear or favour expose before the proper tribunals unprofessional or dishonest conduct by any other member of the profession; and
   (b) undertake only such work as he is competent to perform by virtue of his training and experience, and shall, where advisable, retain and co-operate with other professional engineers or specialists.

Latest Revision Date Not Shown.

THE ASSOCIATION OF PROFESSIONAL ENGINEERS, GEOLOGISTS AND GEOPHYSICISTS OF THE NORTHWEST TERRITORIES

CODE OF ETHICS

1. A Professional Engineer, Geologist or Geophysicist shall recognize that professional ethics is founded upon integrity, competence and devotion to service and to the advancement of human welfare, with due respect to the total environment. This concept shall guide his conduct at all times.

DUTIES OF THE PROFESSIONAL ENGINEER, GEOLOGIST, OR GEOPHYSICIST TO THE PUBLIC

A Professional Engineer, Geologist or Geophysicist:

2. shall have proper regard in all his work for the safety, health and welfare of the public;
3. shall endeavour to extend public understanding of engineering, geology and geophysics and their places in society;
4. shall, where his professional knowledge may benefit the public, seek opportunities to serve in public affairs;
5. shall not be associated with enterprises contrary to the public interest or sponsored by persons of questionable integrity;
6. shall undertake only such work as he is competent to perform by virtue of his training and experience;
7. shall sign and seal only such plans, documents or work as he himself has prepared or carried out or as have been prepared or carried out under his direct professional supervision;
8. shall express opinions on engineering, geological or geophysical matters only on the basis of adequate knowledge and honest conviction.

DUTIES OF THE PROFESSIONAL ENGINEER, GEOLOGIST, OR GEOPHYSICIST TO HIS CLIENT OR EMPLOYER

A Professional Engineer, Geologist or Geophysicist:

9. shall act for his client or employer as a faithful agent or trustee;

10. shall not accept remuneration for services rendered other than from his client or his employer;
11. shall not disclose confidential information without the consent of his client or employer;
12. shall not undertake any assignment which may create a conflict of interest with his client or employer without the full knowledge of the client or employer;
13. shall present clearly to his clients or employers the consequences to be expected if his professional judgment is overruled by other authorities in matters pertaining to work for which he is professionally responsible.

DUTIES OF THE PROFESSIONAL ENGINEER, GEOLOGIST OR GEOPHYSICIST TO THE PROFESSION

A Professional Engineer, Geologist or Geophysicist:

14. shall endeavour at all times to improve the competence, dignity and prestige of his profession;
15. shall conduct himself towards other professional engineers, geologists and geophysicists with fairness and good faith;
16. shall not advertise his professional services in self-laudatory language or in any other manner derogatory to the dignity of his profession;
17. shall not attempt to supplant another engineer, geologist or geophysicist in an engagement after definite steps have been taken toward the other's employment;
18. shall not use the advantages of a salaried position to compete unfairly with another engineer, geologist or geophysicist;
19. shall not exert undue influence or offer, solicit or accept compensation for the purpose of affecting negotiations for an engagement;
20. shall not invite or submit proposals under conditions that constitute price competition for professional services;
21. shall advise the Council of the Association of any practice by another member of his profession which he believes to be contrary to this Code of Ethics.

Latest Revision Not Shown.

ASSOCIATION OF PROFESSIONAL ENGINEERS OF NOVA SCOTIA

CANONS OF ETHICS FOR ENGINEERS

GENERAL

1. A Professional Engineer shall recognize that professional ethics are founded upon integrity, competence and devotion to service and to the advancement of public welfare. This concept shall guide his conduct at all times.

RELATIONS WITH THE PUBLIC

A Professional Engineer:

2. shall regard his duty to public welfare as paramount.
3. shall endeavor to enhance the public regard for his profession by extending the public knowledge thereof.
4. shall undertake only such work as he is competent to perform by virtue of his training and experience.
5. shall sign and seal only such plans, documents or work as he himself has prepared or carried out or as have been prepared or carried out under his direct professional supervision.
6. shall express opinions on engineering matters only on the basis of adequate knowledge, competence and honest conviction.
7. shall express opinions or make statements on engineering projects of public interest that are inspired or paid for by private interest only if he clearly discloses on whose behalf he is giving the opinion or making the statements.
8. shall not be associated with enterprises contrary to public interest or sponsored by persons of questionable integrity.

RELATIONS WITH CLIENTS AND EMPLOYERS

A Professional Engineer:

9. shall act for his client or employer as a faithful agent or trustee and shall act with fairness and justice between his client or employer and the contractor when contracts are involved.

10. shall not accept compensation, financial or otherwise, from more than one interested party for the same service, or for service pertaining to the same work, without the consent of all interested parties.
11. shall not disclose confidential information without the consent of his client or employer.
12. shall not be financially interested in bids on competitive work for which he is employed as an engineer unless he has the consent of his client or employer.
13. shall not undertake any assignment which may create a conflict of interest with his client or employer without the full knowledge of the client or employer.
14. shall present clearly to his clients or employers the consequences to be expected if his professional judgment is overruled by other authorities in matters pertaining to work for which he is professionally responsible.
15. shall refrain from unprofessional conduct or from actions which he considers to be contrary to the public good, even if expected or directed by his employer or client, to act in such a manner.
16. shall not expect or direct an employee or subordinate to act in a manner that he or the employee or subordinate considers to be unprofessional or contrary to the public good.
17. shall guard against conditions that are dangerous or threatening to life, limb or property on work for which he is responsible, or if he is not responsible, will promptly call such conditions to the attention of those who are responsible.

RELATIONS WITH THE PROFESSION

A Professional Engineer:

18. shall co-operate in extending the effectiveness of the engineering profession by interchanging information and experience with other engineers and students and by contributing to the work of engineering societies, schools and the scientific and engineering press.
19. shall endeavour at all times to improve the competence, and thus the dignity and prestige of his profession.
20. shall not advertise his work or merit in a self-laudatory manner and shall avoid all conduct or practice likely to discredit or do injury to the dignity and honor of his profession.
21. shall not attempt to supplant another engineer in an engagement after a definite commitment has been made toward the other's employment.

22. shall not exert undue influence or offer, solicit or accept compensation for the purpose of affecting negotiations for an engagement.
23. shall not compete with another engineer on the basis of charges for work by underbidding, through reducing his normal fees after having been informed of the charges named by the other.
24. shall not use the advantages of a salaried position to compete unfairly with another engineer.
25. shall advise the Discipline Committee of any practice by another member of his profession which he believes to be contrary to this Code of Ethics.
26. shall take care that credit for engineering work is given to those to whom credit is properly due.
27. shall uphold the principle of appropriate and adequate compensation for those engaged in engineering work including those in subordinate capacities as being in the public interest of maintaining the standards of the profession.
28. shall endeavor to provide opportunity for the professional development and advancement of engineers in his employ.

Latest Revision Date Shown 1952.

ASSOCIATION OF PROFESSIONAL ENGINEERS OF ONTARIO

CODE OF ETHICS
from Regulation 538/84
made under the Professional Engineers Act
(Proclaimed in force 1st September 1984)

91. The following is the Code of Ethics of the Association:

1. It is the duty of a practitioner to the public, to his employer, to his clients, to other members of his profession, and to himself to act at all times with,
   i. fairness and loyalty to his associates, employers, clients, subordinates and employees,
   ii. fidelity to public needs, and
   iii. devotion to high ideals of personal honour and professional integrity.
2. A practitioner shall,
   i. regard his duty to public welfare as paramount,
   ii. endeavour at all times to enhance the public regard for his profession by extending the public knowledge thereof and discouraging untrue, unfair or exaggerated statements with respect to professional engineering,
   iii. not express publicly, or while he is serving as a witness before a court, commission or other tribunal, opinions on professional engineering matters that are not founded on adequate knowledge and honest conviction,
   iv. endeavour to keep his licence, temporary licence, limited licence or certificate of authorization, as the case may be, permanently displayed in his place of business.
3. A practitioner shall act in professional engineering matters for each employer as a faithful agent or trustee and shall regard as confidential information obtained by him as to the business affairs, technical methods or processes of an employer and avoid or disclose a conflict of interest that might influence his actions of judgment.
4. A practitioner must disclose immediately to his client any interest, direct or indirect, that might be construed as prejudicial in any way to the professional judgment of the practitioner in rendering service to the client.
5. A practitioner who is an employee-engineer and is contracting in his own name to perform professional engineering work

for other than his employer, must provide his client with a written statement of the nature of his status as an employee and the attendant limitations on his services to the client, must satisfy himself that the work will not conflict with his duty to his employer, and must inform his employer of the work.

6. A practitioner must co-operate in working with other professionals engaged on a project.
7. A practitioner shall,
   - i. conduct himself towards other practitioners with courtesy and good faith,
   - ii. not accept an engagement to review the work of another practitioner for the same employer except with the knowledge of the other practitioner or except where the connection of the other practitioner with the work has been terminated,
   - iii. not maliciously injure the reputation or business of another practitioner,
   - iv. not attempt to gain an advantage over other practitioners by paying or accepting a commission in securing professional engineering work, and
   - v. give proper credit for engineering work, uphold the principle of adequate compensation for engineering work, provide opportunity for professional development and advancement of his associates and subordinates, and extend the effectiveness of the profession through the interchange of engineering information and experience.
8. A practitioner shall maintain the honour and integrity of his profession and without fear or favour expose before the proper tribunals unprofessional, dishonest or unethical conduct by any other practitioner. O.Reg. 538/84, s. 91.

Many of the articles which used to appear in the previous version of the Code of Ethics now appear under section 86 of Regulation 538/84, Professional Misconduct.

PROFESSIONAL MISCONDUCT
from Regulation 538/84

86.—(1) In this section, "negligence" means an act or an omission in the carrying out of the work of a practitioner that constitutes a failure to maintain the standards that a reasonable and prudent practitioner would maintain in the circumstances.

(2) For the purposes of the Act and this Regulation, "professional misconduct" means,

(*a*) negligence;

(*b*) failure to make reasonable provision for the safeguarding of life, health or property of a person who may be affected by the work for which the practitioner is responsible;

(*c*) failure to act to correct or report a situation that the practitioner believes may endanger the safety or the welfare of the public;

(*d*) failure to make responsible provision for complying with applicable statutes, regulations, standards, codes, by-laws and rules in connection with work being undertaken by or under the responsibility of the practitioner;

(*e*) signing or sealing a final drawing, specification, plan, report or other document not actually prepared or checked by the practitioner;

(*f*) failure of a practitioner to present clearly to his employer the consequences to be expected from a deviation proposed in work, if the professional engineering judgment of the practitioner is overruled by non-technical authority in cases where the practitioner is responsible for the technical adequacy of professional engineering work;

(*g*) breach of the Act or regulations, other than an action that is solely a breach of the code of ethics;

(*h*) undertaking work the practitioner is not competent to perform by virtue of his training and experience;

(*i*) failure to make prompt, voluntary and complete disclosure of an interest, direct or indirect, that might in any way be, or be construed as, prejudicial to the professional judgment of the practitioner in rendering service to the public, to an employer or to a client, and in particular without limiting the generality of the foregoing, carrying out any of the following acts without making such a prior disclosure:

1. Accepting compensation in any form for a particular service from more than one party.
2. Submitting a tender or acting as a contractor in respect of work upon which the practitioner may be performing as a professional engineer.
3. Participating in the supply of material or equipment to be used by the employer or client of the practitioner.
4. Contracting in the practitioner's own right to perform professional engineering services for other than the practitioner's employer.
5. Expressing opinions or making statements concerning matters within the practice of professional engineering of

public interest where the opinions or statements are inspired or paid for by other interests;

(*j*) conduct or an act relevant to the practice of professional engineering that, having regard to all the circumstances, would reasonably be regarded by the engineering profession as disgraceful, dishonourable or unprofessional;

(*k*) failure by a practitioner to abide by the terms, conditions or limitations of the practitioner's licence, limited licence, temporary licence or certificate;

(*l*) failure to supply documents or information requested by an investigator acting under section 34 of the Act;

(*m*) permitting, counselling or assisting a person who is not a practitioner to engage in the practice of professional engineering except as provided for in the Act or the regulations. O. Reg. 538/84, s. 86.

ASSOCIATION OF PROFESSIONAL ENGINEERS OF THE PROVINCE OF PRINCE EDWARD ISLAND

CODES OF ETHICS FOR ENGINEERS

FOREWORD

Honesty, justice and courtesy form a moral philosophy which, associated with mutual interest among men, constitutes the foundation of ethics. The engineer should recognize such a standard, not in passive observance, but as a set of dynamic principles guiding his conduct and way of life. It is his duty to practice his profession according to these Canons of Ethics.

As the keystone of professional conduct is integrity the engineer will discharge his duties with fidelity to the public, his employers and clients, and with fairness and impartiality to all. It is his duty to interest himself in public welfare, and to be ready to apply his special knowledge for the benefit of mankind. He should uphold the honor and dignity of his profession and avoid association with any enterprise of questionable character. In his dealings with fellow engineers he should be fair and tolerant.

PROFESSIONAL LIFE

Sec. 1. The engineer will cooperate in extending the effectiveness of the engineering profession by interchanging information and experience with other engineers and students and by contributing to the work of engineering societies, schools and the scientific and engineering press.

Sec. 2. He will not advertise his work or merit in a self-laudatory manner, and he will avoid all conduct or practice likely to discredit or do injury to the dignity and honor of his profession.

RELATIONS WITH THE PUBLIC

Sec. 3. The engineer will endeavour to extend public knowledge of engineering, and will discourage the spreading of untrue, unfair and exaggerated statements regarding engineering.

Sec. 4. He will have due regard for the safety of life and health of public and employees who may be affected by the work for which he is responsible.

Sec. 5. He will express an opinion only when it is founded on adequate knowledge and honest conviction while he is serving as a witness before a court, commission or other tribunal.

Sec. 6. He will not issue ex parte statements, criticisms or arguments on matters connected with public policy which are inspired or paid for by private interests, unless he indicates on whose behalf he is making the statement.

Sec. 7. He will refrain from expressing publicly an opinion on an engineering subject unless he is informed as to the facts relating thereto.

RELATIONS WITH CLIENTS AND EMPLOYERS

Sec. 8. The engineer will act in professional matters for each client or employer as a faithful agent or trustee.

Sec. 9. He will act with fairness and justice between his client or employer and the contractor when dealing with contracts.

Sec. 10. He will make his status clear to his client or employer before undertaking an engagement if he may be called upon to decide on the use of inventions, apparatus, or any other thing in which he may have a financial interest.

Sec. 11. He will guard against conditions that are dangerous or threatening to life, limb or property on work for which he is responsible, or if he is not responsible, will promptly call such conditions to the attention of those who are responsible.

Sec. 12. He will present clearly the consequences to be expected from deviations proposed if his engineering judgment is overruled by non-technical authority in cases where he is responsible for the technical adequacy of engineering work.

Sec. 13. He will engage or advise his client or employer to engage, and he will cooperate with, other experts and specialists whenever the client's or employer's interests are best served by such service.

Sec. 14. He will disclose no information concerning the business affairs or technical process of clients or employers without their consent.

Sec. 15. He will not accept compensation, financial or otherwise, from more than one interested party for the same service, or for services pertaining to the same work, without the consent of all interested parties.

Sec. 16. He will not accept commissions or allowances, directly or indirectly, from contractors or other parties dealing with his client or employer in connection with work for which he is responsible.

Sec. 17. He will not be financially interested in the bids as or of a contractor on competive work for which he is employed as an engineer unless he has the consent of his client or employer.

Sec. 18. He will promptly disclose to his client or employer any interest in a business which may compete with or affect the business of his client or employer. He will not allow an interest in any business to affect his decision regarding engineering work for which he is employed, or which he may be called upon to perform.

RELATIONS WITH ENGINEERS

Sec. 19. The engineer will endeavour to protect the engineering profession collectively and individually from misrepresentation and misunderstanding.

Sec. 20. He will take care that credit for engineering work is given to those to whom credit is properly due.

Sec. 21. He will uphold the principle of appropriate and adequate compensation for those engaged in engineering work, including those in subordinate capacities, as being in the public interest and maintaining the standards of the profession.

Sec. 22. He will endeavour to provide opportunity for the professional development and advancement of engineers in his employ.

Sec. 23. He will not directly or indirectly injure the professional reputation, prospects or practice of another engineer. However, if he considers that an engineer is guilty of unethical, illegal or unfair practice, he will present the information to the proper authority for action.

Sec. 24. He will exercise due restraint in criticizing another engineer's work in public, recognizing the fact that the engineering societies and the engineering press provide the proper forum for technical discussions and criticism.

Sec. 25. He will not try to supplant another engineer in a particular employment after becoming aware that definite steps have been taken toward the other's employment.

Sec. 26. He will not compete with another engineer on the basis of charges for work by underbidding, through reducing his normal fees after having been informed of the charges named by the other.

Sec. 27. He will not use the advantages of a salaried position to compete unfairly with another engineer.

Sec. 28. He will not become associated in responsibility for work with engineers who do not conform to ethical practices.

Latest Revision Date Shown 1955.

ORDRE DES INGÉNIEURS DU QUÉBEC

CODE DE DÉONTOLOGIE DES INGÉNIEURS

SECTION I
DISPOSITIONS GÉNÉRALES

1.01. Le présent règlement est adopté en vertu de l'article 87 du Code des professions (L.R.Q., c. C-26).

1.02. Dans le présent règlement, à moins que le contexte n'indique un sens différent, le mot "client" signifie celui qui bénéficie des services professionnels d'un ingénieur, y compris un employeur.

1.03. La Loi d'interprétation (L.R.Q., c. I-16), avec ses modifications prèsentes et futures, s'applique au présent règlement.

SECTION II
DEVOIRS ET OBLIGATIONS ENVERS LE PUBLIC

2.01. Dans tous les aspects de son travail, l'ingénieur doit respecter ses obligations envers l'homme et tenir compte des conséquences de l'exécution de ses travaux sur l'environnement et sur la vie, la santé et la propriété de toute personne.

2.02. L'ingénieur doit appuyer toute mesure susceptible d'améliorer la qualité et la disponibilité de ses services professionnels.

2.03. L'ingénieur doit, lorsqu'il considère que des travaux sont dangereux pour la sécurité publique, en informer l'Ordre des ingénieurs du Québec ou les responsables de tels travaux.

2.04. L'ingénieur ne doit exprimer son avis sur des questions ayant trait à l'ingéniérie, que si cet avis est basé sur des connaissances suffisantes et sur d'honnêtes convictions.

2.05. L'ingénieur doit favoriser les mesures d'éducation et d'information dans le domaine où il exerce.

SECTION III
DEVOIRS ET OBLIGATIONS ENVERS LE CLIENT
§1. DISPOSITIONS GÉNÉRALES

3.01.01 Avant d'accepter un mandat, l'ingénieur doit tenir compte des limites de ses connaissances et de ses aptitudes ainsi que des moyens dont il peut disposer pour l'exécuter.

3.01.02. L'ingénieur doit reconnaître en tout temps le droit du client de consulter un autre ingénieur. S'il y va de l'intérêt du client, l'ingénieur retient les services d'experts après en avoir informé son client, ou avise ce dernier de le faire.

3.01.03. L'ingénieur doit s'abstenir d'exercer dans des conditions ou des états susceptibles de compromettre la qualité de ses services.

§2. INTÉGRITÉ

3.02.01. L'ingénieur doit s'acquitter de ses obligations professionnelles avec intégrité.

3.02.02. L'ingénieur doit éviter toute fausse représentation concernant sa compétence ou l'efficacité de ses propres services et de ceux généralement assurés par les membres de sa profession.

3.02.03. L'ingénieur doit, dès que possible, informer son client de l'ampleur et des modalités du mandat que ce dernier lui a confié et obtenir son accord à ce sujet.

3.02.04. L'ingénieur doit s'abstenir d'exprimer des avis ou de donner des conseils contradictoires ou incomplets et de présenter ou utiliser des plans, devis et autres documents qu'il sait ambigüs ou qui ne sont pas suffisamment explicites.

3.02.05. L'ingénieur doit informer le plus tôt possible son client de toute erreur préjudiciable et difficilement réparable qu'il a commise dans l'exécution de son mandat.

3.02.06. L'ingénieur doit apporter un soin raisonnable aux biens confiés à sa garde par un client et il ne peut prêter ou utiliser ceux-ci à des fins autres que celles pour lesquelles ils lui ont été confiés.

3.02.07. Si on écarte un avis de l'ingénieur dans le cas où celui-ci est responsable de la qualité technique de travaux d'ingéniérie, l'ingénieur doit indiquer clairement à son client, par écrit, les conséquences qui peuvent en découler.

3.02.08. L'ingénieur ne doit pas recourir à des procédés malhonnêtes ou douteux dans l'exercice de ses activités professionelles.

3.02.09. L'ingénieur doit s'abstenir de verser ou de s'engager à verser, directement ou indirectement, tout avantage, ristourne ou commission en vue d'obtenir un contrat ou lors de l'exécution de travaux d'ingéniérie.

3.02.10. L'ingénieur doit faire preuve d'impartialité dans ses rapports entre son client et les entrepreneurs, fournisseurs et autres personnes faisant affaires avec son client.

§3. DISPONIBILITÉ ET DILIGENCE

3.03.01. L'ingénieur doit faire preuve, dans l'exercice de sa profession, d'une disponibilité et d'une diligence raisonnable.

3.03.02. L'ingénieur doit en plus des avis et des conseils, fournir à son client les explications nécessaires à la compréhension et à l'appréciation des services qu'il lui rend.

3.03.03. L'ingénieur doit rendre compte à son client lorsque celui-ci le requiert.

3.03.04. L'ingénieur ne peut, sauf pour un motif juste et raisonnable, cesser d'agir pour le compte d'un client. Constituent notamment des motifs justes et raisonnables:

a) le fait que l'ingénieur soit en situation de conflit d'intérêts ou dans un contexte tel que son indépendance professionnelle puisse être mise en doute;

b) l'incitation, de la part du client, à l'accomplissement d'actes illégaux, injustes ou frauduleux;

c) le fait que le client ignore les avis de l'ingénieur.

3.03.05. Avant de cesser d'exercer ses fonctions pour le compte d'un client, l'ingénieur doit lui faire parvenir un préavis de délaissement dans un délai raisonnable.

§4. RESPONSABILITÉ

3.04.01. L'ingénieur doit apposer son sceau et sa signature sur l'original et les copies de chaque plans, devis, rapport technique, étude, cahier des charges et autres documents d'ingénierie qu'il a préparés lui-même ou qui ont été préparés sous sa direction et surveillance immédiates par des personnes qui ne sont pas membres de l'Ordre.

L'ingénieur peut également apposer son sceau et sa signature sur l'original et les copies des documents prévus au présent article qui ont été préparés, signés et scellés par un autre ingénieur.

L'ingénieur ne doit ou ne peut apposer son sceau et sa signature que dans les seuls cas prévus au présent article.

§5. INDÉPENDANCE ET DÉSINTÉRESSEMENT

3.05.01. L'ingénieur doit, dans l'exercice de sa profession, subordonner son intérêt personnel à celui de son client.

3.05.02. L'ingénieur doit ignorer toute intervention d'un tiers qui pourrait influer sur l'exécution de ses devoirs professionnels au préjudice de son client.

Sans restreindre la généralité de ce qui précède, l'ingénieur ne doit accepter, directement ou indirectement, aucun avantage ou ristourne en argent ou autrement, d'un fournisseur de marchandises ou de services relativement à des travaux d'ingénierie qu'il effectue pour le compte d'un client.

3.05.03. L'ingénieur doit sauvegarder en tout temps son indépendance professionnelle et éviter toute situation où il serait en conflit d'intérêts.

3.05.04. Dès qu'il constate qu'il se trouve dans une situation de conflit d'intérêts, l'ingénieurs doit en aviser son client et lui demander s'il l'autorise à poursuivre son mandat.

3.05.05. L'ingénieur ne peut partager ses honoraires qu'avec un confrère et dans la mesure où ce partage correspond à une répartition des services et des responsabilités.

3.05.06. L'ingénieur ne doit généralement agir, dans l'exécution d'un mandat, que pour l'une des parties en cause, soit son client. Toutefois, si ses devoirs professionnels exigent qu'il agisse autrement, l'ingénieur doit en informer son client. Il ne doit accepter le versement de ses honoraires que de son client ou du représentant de ce dernier.

§6. SECRET PROFESSIONNEL

3.06.01. L'ingénieur doit respecter le secret de tout renseignement de nature confidentielle obtenu dans l'exercice de sa profession.

3.06.02. L'ingénieur ne peut être relevé du secret professionnel qu'avec l'autorisation de son client ou lorsque la loi l'ordonne.

3.06.03. L'ingénieur ne doit pas faire usage de renseignements de nature confidentielle au préjudice d'un client ou en vue d'obtenir directement ou indirectement un avantage pour lui-même ou pour autrui.

3.06.04. L'ingénieur ne doit pas accepter un mandat qui comporte ou peut comporter la révélation ou l'usage de renseignements ou documents confidentiels obtenus d'un autre client, sans le consentement de ce dernier.

§7. ACCESSIBILITÉ DES DOSSIERS

3.07.01. L'ingénieur doit respecter le droit de son client de prendre connaissance et d'obtenir copie des documents qui le concernent dans tout dossier qu'il a constitué à son sujet.

§8. FIXATION ET PAIEMENT DES HONORAIRES

3.08.01. L'ingénieur doit demander et accepter des honoraires justes et raisonnables.

3.08.02. Les honoraires sont justes et raisonnables s'ils sont justifiés par les circonstances et proportionés aux services rendus. L'ingénieur doit notamment tenir compte des facteurs suivants pour la fixation de ses honoraires:

a) le temps consacré à l'exécution du mandat;

b) la difficulté et l'importance du mandat;

c) la prestation de services inhabituels ou exigeant une compétence ou une célérité exceptionnelles;

d) la responsabilité assumée.

3.08.03. L'ingénieur doit prévenir son client du coût approximatif de ses services et des modalités de paiement. Il doit s'abstenir d'exiger d'avance le paiement de ses honoraires; il peut cependant exiger des avances.

3.08.04. L'ingénieur doit fournir à son client toutes les explications nécessaires à la compréhension de son relevé d'honoraires et des modalités de paiement.

SECTION IV
DEVOIRS ET OBLIGATIONS ENVERS LA PROFESSION
§1. ACTES DÉROGATOIRES

4.01.01. En outre des actes dérogatoires mentionnés aux articles 57 et 58 du Code des professions, est dérogatoire à la dignité de la profession le fait pour un ingénieur:

a) de participer ou de contribuer à l'exercice illégal de la profession;

b) d'inciter quelqu'un de façon pressante ou répétée à recourir à ses services professionnels;

c) de communiquer avec la personne qui a porté plainte sans la permission écrite et préalable du syndic ou de son adjoint, lorsqu'il est informé d'une enquête sur sa conduite ou sa compétence professionnelle ou lorsqu'il a reçu signification d'une plainte à son endroit;

d) de refuser de se soumettre à la procédure de conciliation et d'arbitrage des comptes et à la décision des arbitres;

e) de procéder en justice contre un confrère sur une question relative à l'exercice de la profession avant d'avoir demandé la conciliation au président de l'Ordre;
f) de refuser ou de négliger de se rendre au bureau du syndic, de l'un de ses adjoints ou d'un syndic correspondant, sur demande à cet effet par l'un d'eux;
g) de ne pas avertir le syndic sans délai, s'il croit qu'un ingénieur enfreint le présent règlement.

§2. RELATION AVEC L'ORDRE ET LES CONFRÈRES

4.02.01. L'ingénieur à qui l'Ordre demande de participer à un conseil d'arbitrage de compte, à un comité de discipline ou d'inspection professionnelle, doit accepter cette fonction à moins de motifs exceptionnels.

4.02.02. L'ingénieur doit répondre dans les plus brefs délais à toute correspondance provenant du syndic de l'Ordre, du syndic adjoint ou d'un syndic correspondant, des enquêteurs, des membres du comité d'inspection professionnelle ou du secrétaire de ce dernier comité.

4.02.03. L'ingénieur ne doit pas surprendre la bonne foi d'un confrère, abuser de sa confiance, être déloyal envers lui ou porter malicieusement atteinte à sa réputation. Sans restreindre la généralité de ce qui précède, l'ingénieur ne doit pas notamment:
a) s'attribuer le mérite d'un travail d'ingénierie qui revient à un confrère;
b) profiter de sa qualité d'employeur ou de cadre pour limiter de quelque façon que ce soit l'autonomie professionnelle d'un ingénieur à son emploi ou sous sa responsabilité, notamment à l'égard de l'usage du titre d'ingénieur ou de l'obligation pour tout ingénieur d'engager sa responsabilité professionnelle.

4.02.04. Lorsqu'un client demande à un ingénieur d'examiner ou de réviser des travaux d'ingénierie qu'il n'a pas lui-même exécutés, ce dernier doit en aviser l'ingénieur concerné et, s'il y a lieu, s'assurer que le mandat de son confrère est terminé.

4.02.05. Lorsqu'un ingénieur remplace un confrère dans des travaux d'ingénierie, il doit en avertir ce confrère et s'assurer que le mandat de ce dernier est terminé.

4.02.06. L'ingénieur appelé à collaborer avec un confrère doit préserver son indépendance professionnelle. Si on lui confie une tâche contraire à sa conscience ou à ses principles, il peut demander d'en être dispensé.

§3. CONTRIBUTION À L'AVANCEMENT DE LA PROFESSION

4.03.01. L'ingénieur doit, dans la mesure de ses possibilités, aider au développement de sa profession par l'échange de ses connaissances et de son expérience avec ses confrères et les étudiants, et par sa participation, à titre de professeur ou de maître de stage, aux cours de formation continue et aux stages de perfectionnement.

CODE OF ETHICS OF ENGINEERS

DIVISION I
GENERAL PROVISIONS

1.01. This Regulation is made pursuant to section 87 of the Professional Code (R.S.Q., c. C-26).

1.02. In this Regulation, unless the context indicates otherwise, the word "client" means a person to whom an engineer provides professional services, including an employer.

1.03. The Interpretation Act (R.S.Q., c. I-16), with present and future amendments, applies to this Regulation.

DIVISION II
DUTIES AND OBLIGATIONS TOWARDS THE PUBLIC

2.01. In all aspects of his work, the engineer must respect his obligations towards man and take into account the consequences of the performance of his work on the environment and on the life, health and property of every person.

2.02. The engineer must support every measure likely to improve the quality and availability of his professional services.

2.03. Whenever an engineer considers that certain works are a danger to public safety, he must notify the Ordre des ingénieurs du Québec (Order) or the persons responsible for such work.

2.04. The engineer shall express his opinion on matters dealing with engineering only if such opinion is based on sufficient knowledge and honest convictions.

2.05. The engineer must promote educational and information measures in the field in which he practices.

DIVISION III
DUTIES AND OBLIGATIONS TOWARDS CLIENTS
§1. GENERAL PROVISIONS

3.01.01. Before accepting a mandate, an engineer must bear in mind the extent of his proficiency and aptitudes and also the means at his disposal to carry out the mandate.

3.01.02. An engineer must at all times acknowledge his client's right to consult another engineer. If it is in the client's interest, the engineer shall retain the services of experts after having informed his client thereof, or he shall advise the latter to do so.

3.01.03. An engineer must refrain from practising under conditions or in circumstances which could impair the quality of his services.

§2. INTEGRITY

3.02.01. An engineer must fulfil his professional obligations with integrity.

3.02.02. An engineer must avoid any misrepresentation with respect to his level of competence or the efficiency of his own services and of those generally provided by the members of his profession.

3.02.03. An engineer must, as soon as possible, inform his client of the extent and the terms and conditions of the mandate entrusted to him by the latter and obtain his agreement in that respect.

3.02.04. An engineer must refrain from expressing or giving contradictory or incomplete opinions or advice, and from presenting or using plans, specifications and other documents which he knows to be ambiguous or which are not sufficiently explicit.

3.02.05. An engineer must inform his client as early as possible of any error that might cause the latter prejudice and which cannot be easily rectified, made by him in the carrying out of his mandate.

3.02.06. An engineer must take reasonable care of the property entrusted to his care by a client and he may not lend or use it for purposes other than those for which it has been entrusted to him.

3.02.07. Where an engineer is responsible for the technical quality of engineering work, and his opinion is ignored, the engineer must clearly indicate to his client, in writing, the consequences which may result therefrom.

3.02.08. The engineer shall not resort to dishonest or doubtful practices in the performance of his professional activities.

3.02.09. An engineer shall not pay or undertake to pay, directly or indirectly, any benefit, rebate or commission in order to obtain a contract or upon the carrying out of engineering work.

3.02.10. An engineer must be impartial in his relations between his client and the contractors, suppliers and other persons doing business with his client.

§3. AVAILABILITY AND DILIGENCE

3.03.01. An engineer must show reasonable availability and diligence in the practice of his profession.

3.03.02. In addition to opinion and counsel, the engineer must furnish his client with any explanations necessary to the understanding and appreciation of the services he is providing him.

3.03.03. An engineer must give an accounting to his client when so requested by the latter.

3.03.04. An engineer may not cease to act for the account of a client unless he has just and reasonable grounds for so doing. The following shall, in particular, constitute just and reasonable grounds:

(*a*) the fact that the engineer is placed in a situation of conflict of interest or in a circumstance whereby his professional independence could be called in question;

(*b*) inducement by the client to illegal, unfair or fraudulent acts;

(*c*) the fact that the client ignores the engineer's advice.

3.03.05. Before ceasing to exercise his functions for the account of a client, the engineer must give advance notice of withdrawal within a reasonable time.

§4. RESPONSIBILITY

3.04.01. An engineer must affix his seal and signature on the original and the copies of every plan, specification, technical report, survey, contract specification and other engineering documents prepared by himself or prepared under his immediate control and supervision by persons who are not members of the Order.

An engineer may also affix his seal and signature on the original copies of documents mentioned in this section which have been prepared, signed and sealed by another engineer.

An engineer must not affix his seal and signature except in the cases provided for in this section.

§5. INDEPENDENCE AND IMPARTIALITY

3.05.01. An engineer must, in the practice of his profession, subordinate his personal interest to that of his client.

3.05.02. An engineer must ignore any intervention by a third party which could influence the performance of his professional duties to the detriment of his client.

Without restricting the generality of the foregoing, an engineer shall not accept, directly or indirectly, any benefit or rebate in money or otherwise from a supplier of goods or services relative to engineering work which he performs for the account of a client.

3.05.03. An engineer must safeguard his professional independence at all times and avoid any situation which would put him in conflict of interest.

3.05.04. As soon as he ascertains that he is in a situation of conflict of interest, the engineer must notify his client thereof and ask his authorization to continue his mandate.

3.05.05. An engineer shall share his fees only with a colleague and to the extent where such sharing corresponds to a distribution of services and responsibilities.

3.05.06. In carrying out a mandate, the engineer shall generally act only for one of the parties concerned, namely, his client. However, where his professional duties require that he act otherwise, the engineer must notify his client thereof. He shall accept the payment of his fees only from his client or the latter's representative.

§6. PROFESSIONAL SECRECY

3.06.01. An engineer must respect the secrecy of all confidential information obtained in the practice of his profession.

3.06.02. An engineer shall be released from professional secrecy only with the authorization of his client or whenever so ordered by law.

3.06.03. An engineer shall not make use of confidential information to the prejudice of a client or with a view to deriving, directly or indirectly, an advantage for himself or for another person.

3.06.04. An engineer shall not accept a mandate which entails or may entail the disclosure or use of confidential information or documents obtained from another client without the latter's consent.

§7. ACCESSIBILITY OF RECORDS

3.07.01. An engineer must respect the right of his client to take cognizance of and to obtain copies of the documents that concern the latter in any record which the engineer has made regarding that client.

§8. DETERMINATION AND PAYMENT OF FEES

3.08.01. An engineer must charge and accept fair and reasonable fees.

3.08.02. Fees are considered fair and reasonable when they are justified by the circumstances and correspond to the services rendered. In determining his fees, the engineer must, in particular, take the following factors into account:

(*a*) the time devoted to the carrying out of the mandate;
(*b*) the difficulty and magnitude of the mandate;
(*c*) the performance of unusual services or services requiring exceptional competence or speed;
(*d*) the responsibility assumed.

3.08.03. An engineer must inform his client of the approximate cost of his services and of the terms and conditions of payment. He must refrain from demanding advance payment of his fees; he may, however, demand payment on account.

3.08.04. An engineer must give his client all the necessary explanations for the understanding of his statement of fees and the terms and conditions of its payment.

DIVISION IV
DUTIES AND OBLIGATIONS TOWARDS THE PROFESSION
§1. DEROGATORY ACTS

4.01.01. In addition to those referred to in sections 57 and 58 of the Professional Code, the following acts are derogatory to the dignity of the profession:

(*a*) participating or contributing to the illegal practice of the profession;
(*b*) pressing or repeated inducement to make use of his professional services;
(*c*) communicating with the person who lodged a complaint, without the prior written permission of the syndic or his assistant, whenever he is informed of an inquiry into his professional conduct or competence or whenever a complaint has been laid against him;
(*d*) refusing to comply with the procedures for the conciliation and arbitration of accounts and with the arbitrators' award;
(*e*) taking legal action against a colleague on a matter relative to the practice of the profession before applying for conciliation to the president of the Order;
(*f*) refusing or failing to present himself at the office of the syndic, of one of his assistants or of a corresponding syndic, upon request to that effect by one of those persons;
(*g*) not notifying the syndic without delay if he believes that an engineer infringes this Regulation.

§2. RELATIONS WITH THE ORDER AND COLLEAGUES

4.02.01. An engineer whose participation in a council for the arbitration of accounts, a committee on discipline or a professional inspection committee is requested by the Order, must accept this duty unless he has exceptional grounds for refusing.

4.02.02. An engineer must, within the shortest delay, answer all correspondence addressed to him by the syndic of the Order, the assistant syndic or a corresponding syndic, investigators or members of the professional inspection committee or the secretary of the said committee.

4.02.03. An engineer shall not abuse a colleague's good faith, be guilty of breach of trust or be disloyal towards him or willfully damage his reputation. Without restricting the generality of the foregoing, the engineer shall not, in particular:

(*a*) take upon himself the credit for engineering work which belongs to a colleague;

(*b*) take advantage of his capacity of employer or executive to limit in any way whatsoever the professional autonomy of an engineer employed by him or under his responsibility, in particular with respect to the use of the title of engineer or the obligation of every engineer to be true to his professional responsibility.

4.02.04. Where a client requests an engineer to examine or review engineering work that he has not performed himself, the latter must notify the engineer concerned thereof and, where applicable, ensure that the mandate of his colleague has terminated.

4.02.05. Where an engineer replaces a colleague in engineering work, he must notify that colleague thereof and make sure that the latter's mandate has terminated.

4.02.06. An engineer who is called upon to collaborate with a colleague must retain his professional independence. If a task is entrusted to him and such task goes against his conscience or his principles, he may ask to be excused from doing it.

§3. CONTRIBUTION TO THE ADVANCEMENT OF THE PROFESSION

4.03.01. An engineer must, as far as he is able, contribute to the development of his profession by sharing his knowledge and experience with his colleagues and students, and by his participation as professor or tutor in continuing training periods and refresher training courses.

Latest Revision Date 1983.

ASSOCIATION OF PROFESSIONAL ENGINEERS OF THE PROVINCE OF SASKATCHEWAN

CODE OF ETHICS

For the governance of the conduct of the members and licensees of the Association, the following Code of Ethics is hereby established:

PREAMBLE

Honesty, justice, and courtesy form a moral philosophy which, associated with mutual interest among men, constitute the foundation of ethics. The Professional Engineer should recognize such a standard, not in passive observance but as a set of dynamic principles guiding his conduct and way of life. It is his duty to practise his profession according to this Code of Ethics.

GENERAL

1. A professional engineer owes certain duties to the public, his employers, other members of his profession and to himself and shall act at all times with,
   (a) fidelity to public needs;
   (b) fairness and loyalty to his associates, employers, subordinates and employees; and
   (c) devotion to high ideals of personal honour and professional integrity.

DUTY OF PROFESSIONAL ENGINEER TO THE PUBLIC

2. A professional engineer shall,
   (a) regard his duty to the public as paramount;
   (b) endeavour to maintain public regard for his profession by discouraging untrue, unfair or exaggerated statements with respect to professional engineering;
   (c) not give opinions or make statements on professional engineering projects of public interest that are inspired or paid for

by private interests unless he clearly discloses on whose behalf he is giving the opinions or making the statements;

(d) not express publicly or while he is serving as a witness before a court, commission or other tribunal, opinions on professional engineering matters that are not founded on adequate knowledge and honest conviction;

(e) make effective provisions for the safety of life and health of a person who may be affected by the work for which he is responsible; and shall act to correct or report any situation which he feels may endanger the public;

(f) not knowingly associate with, or allow the use of his name by, an enterprise of doubtful character, nor shall he sanction the use of his reports, in part or in whole, in a manner calculated to mislead, and if it comes to his knowledge they have been so used, shall take immediate steps to correct any false impression given by them;

(g) on all occasions sign or seal reports, plans and specifications which legally require sealing and for which he is professionally responsible;

(h) not offer his services for a fee without first notifying the council of the Association of his intent to do so and of the area of speciality in which he proposes to practise, and receiving from the council of the Association permission to do so.

DUTY OF PROFESSIONAL ENGINEER TO EMPLOYER

3. A professional engineer shall,

(a) act for his employer as a faithful agent or trustee and shall regard as confidential any information obtained by him as to the business affairs, technical methods or processes of his employer, and avoid or disclose any conflict of interest which might influence his actions or judgement;

(b) present clearly to his employers the consequences to be expected from any deviations proposed in the work if his professional engineering judgement is overruled in cases where he is responsible for the technical adequacy of professional engineering work;

(c) advise his employer to engage experts and specialists whenever the employer's interests are best served by so doing;

(d) have no interest, direct or indirect in any materials, supplies or equipment used by his employer or in any persons or firms receiving contracts from his employer unless he informs his employer in advance of the nature of the interest;

(e) not act as consulting engineer in respect of any work upon which he may be the contractor unless he first advises his employer; and

(f) not accept compensation, financial or otherwise, for a particular service, from more than one person except with the full knowledge of all interested parties.

DUTY OF PROFESSIONAL ENGINEER TO OTHER PROFESSIONAL ENGINEERS

4. A Professional Engineer shall,

(a) not attempt to supplant another engineer after definite steps have been taken towards the other's employment;

(b) not accept employment by a client, knowing that a claim for compensation or damages, or both, of a fellow engineer previously employed by the same client, and whose employment has been terminated, remains unsatisfied, or until such claim has been referred to arbitration or issue has been joined at law, or unless the engineer previously employed has neglected to press his claim legally, or the council of the Association gives its consent;

(c) not accept any engagement to review the work of another professional engineer for the same employer except with the knowledge of that engineer, or except where the connection of that engineer with the work has been terminated;

(d) not maliciously injure the reputation or business of another professional engineer;

(e) not attempt to gain an advantage over other members of his profession by paying or accepting a commission in securing professional engineering work, or by reducing his fees below the approved minimums;

(f) not advertise in a misleading manner or in a manner injurious to the dignity of his profession;

(g) give proper credit for engineering work;

(h) uphold the principle of adequate compensation for engineering work;

(i) provide opportunity for professional development and advancement of his professional colleagues;

(j) extend the effectiveness of the profession through the interchange of engineering information and experiences.

DUTY OF PROFESSIONAL ENGINEER TO HIMSELF

5. A Professional Engineer shall,
   (a) maintain the honour and integrity of his profession and without fear or favour expose before the proper tribunals unprofessional or dishonest conduct by any other member of the profession;
   (b) undertake only such work as he is competent to perform by virtue of his training and experience; and
   (c) constantly strive to broaden his knowledge and experience by keeping abreast of new techniques and developments in his field of endeavour.

COMING INTO EFFECT

38. These bylaws shall come into effect on the 30th day of November 1961.

Latest Revision Date Shown 1982.

THE ASSOCIATION OF PROFESSIONAL ENGINEERS OF THE YUKON TERRITORY

BY-LAWS

---

16. Code of Ethics
    (a) General
    A professional engineer owes certain duties to the public, to his employers, to his clients, to other members of his profession and to himself and shall act at all times with:
    1. Fairness and loyalty to his associates, employers, clients, subordinates and employees;
    2. fidelity to public needs and
    3. devotion to high ideals of personal honour and professional integrity.

    (b) Duty of Professional Engineer to the Public
    A professional engineer shall:
    1. regard his duty to public welfare as paramount;
    2. endeavour at all times to enhance the public regard for his profession by extending the public knowledge thereof and discouraging untrue, unfair or exaggerated statements with respect to professional engineering;
    3. not give opinions or make statements, on professional engineering projects of public interest, that are inspired or paid for by private interests unless he clearly discloses on whose behalf he is giving the opinions or making the statements;
    4. not express publicly, or while he is serving as a witness before a court, commission or other tribunal, opinions on professional engineering matters that are not founded on adequate knowledge and honest conviction;
    5. make effective provisions for the safety of life and health of a person who may be affected by the work for which he is responsible; and at all times shall act to correct or report any situation which he feels may endanger the safety or welfare of the public;
    6. make effective provisions for meeting lawful standards, rules, or regulations relating to environmental control and protection in connection with any work being undertaken by him or under his responsibility;
    7. sign or seal only those plans, specifications and reports made by him or under his personal supervision and direction or those which have been thoroughly reviewed by him as if they were his own work, and found to be satis-

factory and

8. refrain from associating himself with or allowing the use of his name by an enterprise of questionable character.

(c) Duty of Professional Engineer to Employer

A professional engineer shall:

1. act in professional engineering matters for each employer as a faithful agent or trustee and shall regard as confidential any information obtained by him as to the business affairs, technical methods or processes of an employer and avoid or disclose any conflict of interest which might influence his actions or judgement;
2. present clearly to his employers the consequences to be expected from any deviations proposed in the work if he is informed that his professional engineering judgement is overruled by nontechnical authority in cases where he is responsible for the technical adequacy of professional engineering work;
3. have no interest, direct or indirect, in any materials, supplies or equipment used by his employer or in any persons or firms receiving contracts from his employer unless he informs his employer in advance of the nature of the interest;
4. not tender on competitive work upon which he may be acting as a professional engineer unless he first advises his employer;
5. not act as consulting engineer in respect of any work upon which he may be the contractor unless he first advises his employer, and
6. not accept compensation, financial or otherwise for a particular service from more than one person except with the full knowledge of all interested parties.

(d) Duty of Professional Engineer in Independent Practice to Client

A professional engineer in private practice, in addition to all other sections, shall:

1. disclose immediately any interest, direct or indirect, which may in any way be constituted as prejudicial to his professional judgement in rendering service to his client;
2. if he is an employee-engineer and is contracting in his own name to perform professional engineering work for other than his employer, clearly advise his client as to the nature of his status as an employee and the attendant limitations on his services to the client. In addition he shall ensure that such work will not conflict with his duty to his employer;

3. carry out his work in accordance with applicable statutes, regulations, standards, codes, and by-laws; and
4. co-operate as necessary in working with such other professionals as may be engaged on a project.

(e) Duty of Professional Engineer to Other Professional Engineers

A professional engineer shall:

1. conduct himself towards other professional engineers with courtesy and good faith;
2. not accept any engagement to review the work of another professional engineer for the same employer or client except with the knowledge of that engineer, or except where the connection of that engineer with the work has been terminated;
3. not maliciously injure the reputation or business of another professional engineer;
4. not attempt to gain an advantage over other members of his profession by paying or accepting a commission in securing professional engineering work;
5. not advertise in a misleading manner or in a manner injurious to the dignity of his profession, but shall seek to advertise by establishing a well-merited reputation for personal capability; and
6. give proper credit for engineering work, uphold the principle of adequate compensation for engineering work, provide opportunity for professional development and advancement of his associates and subordinates; and extend the effectiveness of the profession through the interchange of engineering information and experience.

(f) Duty of Professional Engineer to Himself

A professional engineer shall:

1. maintain the honour and integrity of his profession and without fear or favour expose before the proper tribunals unprofessional or dishonest conduct by any other members of the profession; and
2. undertake only such work as he is competent to perform by virtue and training and experience, and shall, where advisable, retain and co-operate with other professional engineers or specialists.

17. Repeal of Old By-Laws

Upon the coming into force of the foregoing By-Laws, all the By-Laws of the Association previously in force shall stand revoked.

Latest Revision Date Shown 1981.

CUT HERE

STUDENT REPLY CARD

In order to improve future editions, we are seeking your comments on *Professional Engineering Practice: Ethical Aspects, third edition,* by Morrison and Hughes.

After you have read this text, please answer the following questions and return this form via Business Reply Mail. *Thanks in advance for your feedback!*

1. Name of your college or university: ______________________

2. Major program of study: ______________________

3. Your instructor for this course: ______________________

4. Are there any sections of this text which were not assigned as course reading? If so, please specify those chapters or portions: ____________

5. How would you rate the overall accessibility of the content? Please feel free to comment on reading level, writing style, terminology, layout and design features, and such learning aids as chapter objectives, summaries, and appendices.

CUT HERE

FOLD HERE

6. What did you like *best* about this book?

7. What did you like *least?*

If you would like to say more, we'd love to hear from you. Please write to us at the address shown on the reverse of this card.

CUT HERE

FOLD HERE

BUSINESS REPLY MAIL

No Postage Stamp Necessary If Mailed in Canada

Postage will be paid by

Attn: Sponsoring Editor, Engineering
The College Division
McGraw-Hill Ryerson Limited
300 Water Street
Whitby, Ontario
L1N 9Z9

TAPE SHUT